Universitext

For other titles published in this series, go to
www.springer.com/series/223

Anatoli Andrianov

Introduction to Siegel Modular Forms and Dirichlet Series

Springer

Anatoli Andrianov
Russian Academy of Sciences
Steklov Institute of Mathematics
Fontanka 27
191023 St. Petersburg
Russia
anandr@pdmi.ras.ru

ISBN 978-0-387-78752-7 ISBN 978-0-387-78753-4 (eBook)
DOI 10.1007/978-0-387-78753-4

Library of Congress Control Number: 2008938066

Mathematics Subject Classification (2000): 11Fxx, 11F66

This is a translation of the Dutch, *Meetkunde*, originally published by Epsilon–Uitgaven, 2000.

Printed on acid-free paper

springer.com

To Goro Shimura and to my granddaughter Sasha

Preface

Several years ago I was invited to an American university to give one-term graduate course on Siegel modular forms, Hecke operators, and related zeta functions. The idea to present in a concise but basically complete and self-contained form an introduction to an important and developing area based partly on my own work attracted me. I accepted the invitation and started to prepare the course. Unfortunately, the visit was not realized. But the idea of such a course continued to be alive till after a number of years this book was finally completed. I hope that this short book will serve to attract young researchers to this beautiful field, and that it will simplify and make more pleasant the initial steps.

No special knowledge is presupposed for reading this book beyond standard courses in algebra and calculus (one and several variables), although some skill in working with mathematical texts would be helpful. The reader will judge whether the result was worth the effort.

Dedications. The ideas of Goro Shimura exerted a deep influence on the number theory of the second half of the twentieth century in general and on the author's formation in particular.

When Andrè Weil was signing a copy of his "Basic Number Theory" to my son, he wrote in Russian, "To Fedor Anatolievich hoping that he will become a number theoretist". Fedor has chosen computer science. Now I pass on the idea to Fedor's daughter, Alexandra Fedorovna.

Contents. The main objective of this book is to give a concise but basically complete and self-contained introduction to the multiplicative theory of Siegel modular forms, Hecke operators, and zeta functions, including the classical case of modular forms in one variable. Chapter 1 contains a compressed exposition of essential features of the theory of Siegel modular forms of integral weight for congruence subgroups of the symplectic modular group $\mathrm{Sp}_n(\mathbb{Z})$ of arbitrary genus n. Chapter 2 treats analytical properties of radial Dirichlet series attached to modular forms of genera 1 and 2. Chapter 3 is dealing with the abstract theory of Hecke–Shimura rings for symplectic and related group. Action of Hecke operators on Siegel modular forms is considered in Chapter 4. In Chapter 5, we examine applications of

Hecke operators to a study of multiplicative properties of Fourier coefficients of modular forms and the related Euler product factorization of radial Dirichlet series attached to eigenfunctions. This leads us to Hecke zeta functions of modular forms in one variable and to spinor (or Andrianov) zeta functions of Siegel modular forms of genus two. At the end of this chapter we arrive at the proof of the analytic continuation and functional equation (under certain assumptions) of Euler products associated with modular forms of genus two.

The book contains a number of exercises that usually consider some interesting points not included in the main text, partly for the reasons of space and partly because of their special character.

References. I try to present the proofs in full detail whenever it is reasonable and possible. The main text contains no references. Essential references are included in the **Notes** at the end of the book. It should be noted that the author does not pretend to give an encyclopedic survey of modular forms or a complete bibliography, but rather to hint at certain principal points of the theory and illustrate how they have been reached.

Acknowledgments. I would like to express my deep gratitude to Fedor Andrianov, the assumed co–author, who carefully read the manuscript and suggested a number of improvements. I am very grateful to Mark Spencer, of Springer New York for continued interest in this book and highly effective cooperation. Great acknowledgment must also go to Charlene Cruz Cedras, the editorial assistant of Mark Spencer, who has given me valuable assistance and support. Finally, I must acknowledge my grateful thanks to Dr. David Kramer for highly careful editing of the manuscript, that I have never encountered.

St Petersburg *Anatoli Andrianov*
January/April 2008

Contents

General Notation

The letters $\mathbb{N}$, $\mathbb{P}$, $\mathbb{Z}$, $\mathbb{Q}$, $\mathbb{R}$, and $\mathbb{C}$ are reserved for the set of positive rational integers, the set of positive rational prime numbers, the ring of rational integers, the field of rational numbers, the field of real numbers, and the field of complex numbers, respectively.

$K[x_1,\dots,x_n]$ is the ring of polynomials in $x_1,\dots,x_n$ with coefficients in K, and $\mathbb{A}_n^m$ is the set of all $m \times n-$matrices with entries in a set $\mathbb{A}$.

If M is a matrix, ${}^tM$ always denotes the transpose of M, $\sigma(M)$ for a square M is the trace of M, and $\overline{M}$ for a complex matrix M means the matrix with conjugate entries. If Y is a real symmetric matrix, then $Y > 0$ (resp., $Y \geqq 0$) means that Y is positive definite (resp., positive semidefinite). For two matrices A and B of suitable dimensions we write

$$A[B] = {}^tBAB.$$

Introduction: The Two Features of Arithmetic Zeta Functions

A *zeta function* in arithmetic is, generally speaking, a generating function for an arithmetic problem written in the form of a Dirichlet series. A well stated zeta function must have at least two principal features: an *Euler product factorization* and an *analytic continuation* over the whole complex plane satisfying *functional equations.* The first reflects relations between the global arithmetic problem and its localizations, while the second provides a kind of reciprocity between the localizations.

Let us illustrate this with some examples

Riemann Zeta Function. The *Riemann zeta function,*

$$\zeta(s) = \sum_{n=1}^{\infty} \frac{1}{n^s} \qquad (\mathrm{Re}\, s > 1),$$

is the generating function for the numbers of ideals of given norm in the ring $\mathbb{Z}$ of rational integers. It has an Euler product factorization of the form

$$\zeta(s) = \prod_{p} \left(1 - \frac{1}{p^s}\right)^{-1} \qquad (\mathrm{Re}\, s > 1),$$

the product being taken over all rational prime numbers p.

It was Bernhard Riemann who proved in the middle of the nineteenth century that $\zeta(s)$ has an analytic continuation over the whole complex s-plane, is holomorphic except for a simple pole of residue 1 at $s = 1$, and satisfies the functional equation that the function $\pi^{-s/2}\Gamma(s/2)\zeta(s)$, where Γ is the gamma function, is invariant under the substitution $s \mapsto 1 - s$. He also discovered that the problem of *distribution of prime numbers* is closely connected with the *location of complex zeros* of the zeta function in the vertical strip $0 \leq \Re s \leq 1$. At the end of the century, J. Hádamard and Ch. de la Vallée Poussin proved that $\zeta(s)$ has no zeros on the line $\Re s = 1$, which implied the famous asymptotic formula for the number $\pi(x)$ of prime numbers not exceeding x,

$$\pi(x) \sim \frac{x}{\log x} \qquad (x \to \infty).$$

Zeta Function of Algebraic Varieties. The *global zeta function of a nonsingular algebraic variety* V over the field $\mathbb{Q}$ of rational numbers is defined by an Euler product

$$\zeta^*(V,s) = \prod_p \zeta(V_p, p^{-s})$$

of *local zeta functions* $\zeta(V_p, p^{-s})$, where p runs over all prime numbers such that V has a "good" and, in particular, nonsingular *reduction* V_p *modulo* p, i.e., the variety over the finite field $\mathbb{F}(p)$ of p elements obtained by replacing equations defining V with corresponding congruences modulo p. The local zeta function is the zeta function of V_p defined by

$$\zeta(V_p,t) = \exp\left(\sum_{\delta=1}^{\infty} N(p^\delta)t^\delta/\delta\right),$$

where $N(p^\delta)$ is the number of points on V_p with coordinates in the finite field $\mathbb{F}(p^\delta)$ of p^δ elements. According to Bernhard Dwork, the local zeta functions $\zeta(V_p,t)$ of nonsingular varieties V_p are rational fractions in t. It follows that the global zeta function $\zeta^*(V,s)$ can be written as a Dirichlet series convergent in a right half-plane of the complex variable s. It is generally believed that the zeta function can be analytically continued over the whole $s-$plane as a meromorphic function and satisfies functional equations, but it is doubtful that a human being living now will see a complete proof.

Nevertheless, even particular cases present considerable interest (all genuine number theory consists of particular cases; everything else is algebra). Let us consider a *(projective) elliptic curve*

$$E: \quad y^2z = x^3 + axz^2 + bz^3 \qquad (a,b \in \mathbb{Z}).$$

The points on E with coordinates in $\mathbb{Q}$ form an abelian group, which we denote by $E_{\mathbb{Q}}$; a theorem of L.J. Mordell tells us that the group $E_{\mathbb{Q}}$ is finitely generated, i.e., is a product of a finite group by a lattice of finite rank g. A principal problem of the theory is to determine the group $E_{\mathbb{Q}}$, and, in particular, to determine the rank g. In the mid 1960's B.J. Birch and H.P.F. Swinnerton-Dyer put forward revolutionary conjectures connecting the group $E_{\mathbb{Q}}$ with the zeta function $\zeta^*(E,s)$ of the curve. Let us recall some details.

A prime number p is said to be good if it does not divide $6(27b^2+4a^3)$. For such a prime p, the reduction

$$E_p: \quad y^2z \equiv x^3 + axz^2 + bz^3 \pmod{p}$$

of E modulo p is an elliptic curve over $\mathbb{F}(p)$. It is well known that the zeta function of E_p over $\mathbb{F}(p)$ has the form

$$\zeta(E_p,t) = \frac{1-(1+p-N(p))t+pt^2}{(1-t)(1-pt)}.$$

Then we may define a zeta function of E by

$$\zeta^*(E,s) = \prod_p \left(1-(1+p-N(p))p^{-s}+p^{1-2s}\right)^{-1},$$

where the product is taken over all good primes. It converges for $\Re s > 3/2$. Then the main *Birch–Swinnerton-Dyer conjecture* says that $\zeta^*(E,s)$ has a zero of order g at $s=1$. Generally, it is still open.

Zeta Functions of Automorphic Forms. Despite the clear importance of zeta functions of algebraic varieties, algebraic geometry provides no means for their investigation. The only hope is to relate them to techniques coming from an analytic background, probably with *zeta functions of automorphic forms*, which, in contrast, usually have vast means for analytic investigation, but often lack clear arithmetic motivation.

Let us consider the simplest case of a Dirichlet series of modular forms of integral weight for congruence subgroups K of the modular group $\mathrm{SL}_2(\mathbb{Z})$. Let us recall the corresponding definitions.

A function F on the upper half-plane $\mathbb{H} = \{x+iy \in \mathbb{C} \mid y > 0\}$ is said to be a *modular cusp form* of weight k for the group K if it is holomorphic on $\mathbb{H}$, equals zero at all cusps of K, and satisfies

$$(cz+d)^{-k} F\left(\frac{az+b}{ac+d}\right) = F(z) \quad \text{for each} \quad \begin{pmatrix} a & b \\ c & d \end{pmatrix} \in K.$$

All such functions form a finite dimensional space $\mathfrak{N} = \mathfrak{N}_k(K)$ over the field $\mathbb{C}$ of complex numbers. If the group K contains the matrix $\left(\begin{smallmatrix} 1 & 1 \\ 0 & 1 \end{smallmatrix}\right)$, then every $F \in \mathfrak{N}$ can be presented by a *Fourier series* absolutely convergent on $\mathbb{H}$ of the form

$$F(z) = \sum_{m=1}^{\infty} f(m) e^{2\pi i m z}$$

with constant Fourier coefficients $f(m)$. Let us associate with F the *Dirichlet series of F* defined by

$$Z(F,s) = \sum_{m=1}^{\infty} \frac{f(m)}{m^s}.$$

The series converges absolutely in a right half-plane of the variable s, and can be presented there by means of a Mellin integral

$$\Phi(s) = \Phi(F,s) = \int_0^{\infty} F(iy) y^{s-1} dy = (2\pi)^{-s}\Gamma(s) Z(F,s).$$

Suppose that the group K satisfies

$$\begin{pmatrix} 0 & -1 \\ q & 0 \end{pmatrix}^{-1} K \begin{pmatrix} 0 & -1 \\ q & 0 \end{pmatrix} = K$$

for a positive integer q. Then it is easy to check that for each cusp form $F \in \mathfrak{N}$, the function

$$(F \mid \omega)(z) = q^{-k/2} z^{-k} F(-1/qz)$$

again belongs to $\mathfrak{N}$ and satisfies $F \mid \omega \mid \omega = (-1)^k F$. It follows that one can write the direct sum decomposition

$$\mathfrak{N} = \mathfrak{N}^+ + \mathfrak{N}^-, \quad \text{where for} \quad F \in \mathfrak{N}^\pm, \quad F \mid \omega = \pm i^k F.$$

Exercise. Prove the above assertions.

If $F \in \mathfrak{N}^\pm$, then

$$F(i/qy) = \pm(-1)^k q^{k/2} y^k F(iy) \qquad (y > 0),$$

and we can write, for $\Re s$ sufficiently large, that

$$\begin{aligned}
\Phi(s) &= \int_0^{q^{-1/2}} F(iy) y^{s-1} dy + \int_{q^{-1/2}}^{\infty} F(iy) y^{s-1} dy \\
&= \int_{q^{-1/2}}^{\infty} F(i/qy)(1/qy)^{s-1}(1/qy^2) dy + \int_{q^{-1/2}}^{\infty} F(iy) y^{s-1} dy \\
&= \pm(-1)^k q^{k/2-s} \int_{q^{-1/2}}^{\infty} F(iy) y^{k-s-1} dy + \int_{q^{-1/2}}^{\infty} F(iy) y^{s-1} dy.
\end{aligned}$$

Both of the last integrals are holomorphic for all s, and so is the function $\Phi(s)$. Moreover, the last expression implies that

$$\Phi(k-s) = \pm(-1)^k q^{s-k/2} \Phi(s),$$

which is the functional equation for the Dirichlet series $Z(F, s)$.

Note that the simplest of the groups K satisfying the given conditions is the group

$$\Gamma_0(q) = \left\{ \begin{pmatrix} a & b \\ c & d \end{pmatrix} \in \mathrm{SL}_2(\mathbb{Z}) \;\middle|\; c \equiv 0 \pmod{q} \right\}.$$

The problem of Euler product factorization of the Dirichlet series corresponding to modular forms of integral weight for the groups of the type $\Gamma_0(q)$ was essentially solved by Erich Hecke in 1937 and completed by A.O.L. Atkin and J. Lehner in 1970. In particular, it was found that although the Dirichlet series of a cusp form does not necessarily have an Euler product factorization, the space of cusp forms has a basis consisting of forms with Dirichlet series decomposable into Euler products, which we call the *zeta functions of modular forms*. Such forms can be characterized as eigenfunctions of certain rings of linear operators, the Hecke operators, acting on the spaces $\mathfrak{N}_k(K)$.

Since the nineteenth century, the main arithmetic application of modular forms had been the analytical theory of integral quadratic forms. The reason is that the generating Fourier series with coefficients equal to numbers of integral representations of positive integers by a positive definite integral quadratic form is a modular (not cusp) form. But in the middle of the twentieth century Goro Shimura and Yutaka Taniyama proposed famous conjectures relating modular forms and elliptic curves over $\mathbb{Q}$. The *Shimura–Taniyama conjecture* includes the conjecture that the zeta function $\zeta^*(E, s)$ of every elliptic curve E over $\mathbb{Q}$ completed by appropriate p-factors for bad primes is, in fact, the zeta function of a cusp form of weight 2 for the group $\Gamma_0(q)$, where q is the product of some degrees of bad primes. In 1985, G. Frey made the remarkable observation that this conjecture would imply Fermat's last theorem. The precise relation of the two was established later by K.A. Ribet, which allowed A. Wiles in 1995 to prove the Fermat's last theorem, one of the brightest achievements of mathematics of the twentieth century.

One can hardly doubt that the relation between zeta functions of elliptic curves and zeta functions of modular forms in one variable described by the Shimura–Taniyama conjecture is only a particular case of some general links between global zeta functions of algebraic varieties and zeta functions of automorphic forms. Speaking of abelian varieties in place of elliptic curves, one can expect that modular forms in one variable should be replaced by Siegel modular forms for congruence subgroups of the symplectic modular group $\Gamma^n = \mathrm{Sp}_n(\mathbb{Z})$, and this expectation is supported by some numerical evidence.

Chapter 1
Modular Forms

1.1 The Symplectic Group and the Upper Half-Plane

Symplectic Matrices. A matrix $M \in \mathbb{C}_{2n}^{2n}$ is said to be *symplectic* if it satisfies the relation

$$ {}^tMJM = \mu(M)J \qquad \text{with} \quad J = J_n = \begin{pmatrix} 0 & 1_n \\ -1_n & 0 \end{pmatrix} \tag{1.0}$$

and a nonzero scalar $\mu(M)$ called the *multiplier* of M, where $0 = 0_n$ and 1_n are the zero matrix and the unit matrix of order n, respectively. It is clear that the product of two symplectic matrices of the same order is again a symplectic matrix, and the multiplier of a product is the product of multipliers of factors.

Lemma 1.1. *Let $M = \left(\begin{smallmatrix} A & B \\ C & D \end{smallmatrix}\right)$, where the blocks A, B, C, and D are complex square matrices of order n, and let μ be a nonzero complex number. Then the following conditions are equivalent:*

(1) *M is symplectic with the multiplier $\mu(M) = \mu$;*
(2) *${}^tM$ is symplectic with the multiplier $\mu({}^tM) = \mu$;*
(3) *M is invertible, and*

$$\mu M^{-1} = \begin{pmatrix} {}^tD & -{}^tB \\ -{}^tC & {}^tA \end{pmatrix}; \tag{1.1}$$

(4) *the blocks A, B, C, and D satisfy the conditions*

$$ {}^tAC = {}^tCA, \quad {}^tBD = {}^tDB, \quad \text{and} \quad {}^tAD - {}^tCB = \mu 1_n \tag{1.2}$$

or the conditions

$$ A{}^tB = B{}^tA, \quad C{}^tD = D{}^tC, \quad \text{and} \quad A{}^tD - B{}^tC = \mu 1_n. \tag{1.3}$$

Proof. This is an easy but useful exercise on multiplication of block–matrices, which we leave to the reader. □

A. Andrianov, *Introduction to Siegel Modular Forms and Dirichlet Series*, Universitext,
DOI 10.1007/978-0-387-78753-4_1,

Exercise 1.2. Prove the lemma.

In the course of our arithmetic considerations we shall be interested in discrete subgroups and subsemigroups of the *general real positive symplectic group of genus n* consisting of all real symplectic matrices of order $2n$ with positive multipliers:

$$\mathbb{G} = \mathbb{G}^n = \mathrm{GSp}_n^+(\mathbb{R}) = \left\{ M \in \mathbb{R}_{2n}^{2n} \middle| \quad {}^tMJM = \mu(M)J, \, \mu(M) > 0 \right\}. \tag{1.4}$$

Action on Upper Half-Plane. The group $\mathbb{G}$ is a real Lie group acting as a group of analytic automorphisms on the $n(n+1)/2-$dimensional open complex variety

$$\mathbb{H} = \mathbb{H}^n = \left\{ Z = X + iY \in \mathbb{C}_n^n \middle| \quad {}^tZ = Z, \quad Y > 0 \right\}, \tag{1.5}$$

called the *upper half-plane of genus n*, with the action defined by the rule

$$\mathbb{G} \ni M = \begin{pmatrix} A & B \\ C & D \end{pmatrix} : Z \mapsto M\langle Z\rangle = (AZ+B)(CZ+D)^{-1} \qquad (Z \in \mathbb{H}). \tag{1.6}$$

In order to verify this, we have to check that the mapping (1.6) is always defined, maps the upper half-plane into itself, and satisfies

$$(MM')\langle Z\rangle = M\langle M'\langle Z\rangle\rangle \qquad (M, M' \in \mathbb{G}, Z \in \mathbb{H}). \tag{1.7}$$

The relations (1.7) follow from the definition by a formal comparison of both parts, provided that both parts are defined. So it would be sufficient to prove the following two lemmas.

Lemma 1.3. *For every matrices* $M = \left(\begin{smallmatrix} A & B \\ C & D \end{smallmatrix}\right) \in \mathbb{G}$ *and* $Z \in \mathbb{H}$, *the matrices*

$$J(M, Z) = CZ + D \tag{1.8}$$

are nonsingular and satisfy the rule

$$J(MM', Z) = J(M, M'\langle Z\rangle)J(M', Z) \qquad (M, M' \in \mathbb{G}, Z \in \mathbb{H}). \tag{1.9}$$

Proof. The relation (1.9) formally follows from the definitions if the matrix $J(M', Z)$ is nonsingular.

As to nonsingularity, we note first that the matrix (1.8) is nonsingular for every $M = \left(\begin{smallmatrix} A & B \\ C & D \end{smallmatrix}\right) \in \mathbb{G}$ and $Z = i1_n$. Indeed, otherwise, the matrix

$$(Ci+D)^t\overline{(Ci+D)} = C^tC + D^tD + i(C^tD - D^tC) = C^tC + D^tD$$

(see (1.3)) is singular, and so since it is symmetric and semidefinite, there is a nonzero real n-column T such that ${}^tT(C^tC + D^tD)T = 0$, whence ${}^tTC^tCT = 0$ and ${}^tTD^tDT = 0$, and so ${}^tTC = {}^tTD = 0$. The last relations imply that the rank of the matrix (C, D) is less than n, which is impossible, since the matrix M is nonsingular.

Now note that each matrix $Z = X + iY \in \mathbb{H}$ can be written in the form $Z = M'\langle i1_n \rangle$ with a matrix $M' \in \mathbb{G}$. It is sufficient to take $M' = \begin{pmatrix} A_1 & X{}^tA_1^{-1} \\ 0 & {}^tA_1^{-1} \end{pmatrix}$, if $Y = A_1{}^tA_1$. Finally, by (1.9) with $Z = i1_n$, we get

$$J(MM', i1_n) = J(M, M'\langle i1_n \rangle)J(M', i1_n) = J(M, Z)J(M', i1_n),$$

which implies that $J(M,Z)$ is nonsingular. □

Lemma 1.4. *Let $M = \left(\begin{smallmatrix} A & B \\ C & D \end{smallmatrix}\right) \in \mathbb{G}$ and $Z = X + iY \in \mathbb{H}$ then the matrix $Z' = X' + iY' = M\langle Z \rangle$ is symmetric, and*

$$Y' = \mu(M){}^t(C\overline{Z} + D)^{-1}Y(CZ + D)^{-1}. \tag{1.10}$$

In particular, $Z' \in \mathbb{H}$.

Proof. As follows easily from the relations (1.2), the matrix

$${}^t(CZ + D)M\langle Z \rangle(CZ + D) = (Z{}^tC + {}^tD)(AZ + B) = Z{}^tCAZ + {}^tDAZ + Z{}^tCB + {}^tDB$$

is symmetric, and so the matrix $Z' = M\langle Z \rangle$ is symmetric too. Further, by (1.2), we have

$$\begin{aligned} {}^t(C\overline{Z} + D)(Z' - \overline{Z}')(CZ + D) &= (\overline{Z}{}^tC + {}^tD)(AZ + B) - (\overline{Z}{}^tA + {}^tB)(CZ + D) \\ &= \overline{Z}({}^tCA - {}^tAC)Z + ({}^tDA - {}^tBC)Z + \overline{Z}({}^tCB - {}^tAD) \\ &= \mu(M)(Z - \overline{Z}). \end{aligned}$$

The formula (1.10) and the lemma follow. □

The above formulas will allow us to find a $\mathbb{G}$-invariant element of volume on $\mathbb{H}$. The upper half-plane $\mathbb{H}$ is clearly an open subset of the $n(n+1)$-dimensional real affine space, and we can consider the Euclidean element of volume on $\mathbb{H}$,

$$dZ = \prod_{1 \le \alpha \le \beta \le n} dx_{\alpha\beta}dy_{\alpha\beta} \qquad (Z = (x_{\alpha\beta} + iy_{\alpha\beta}) \in \mathbb{H}). \tag{1.11}$$

Lemma 1.5. *For each matrix $M = \left(\begin{smallmatrix} A & B \\ C & D \end{smallmatrix}\right) \in \mathbb{G}$, the element of volume* (1.11) *satisfies the relation*

$$dM\langle Z \rangle = \mu(M)^{n(n+1)}|\det(CZ + D)|^{-2n-2}dZ.$$

Proof. For $Z = (z_{\alpha\beta}) = (x_{\alpha\beta} + iy_{\alpha\beta}) \in \mathbb{H}$, we set $Z' = (z'_{\gamma\delta}) = (x'_{\gamma\delta} + iy'_{\gamma\delta}) = M\langle Z \rangle$. We have to compute the absolute value of the Jacobian of the variables $x'_{\gamma\delta}, y'_{\gamma\delta}$ with respect to the variables $x_{\alpha\beta}, y_{\alpha\beta}$. First, we shall consider the transformation of the differentials of the complex variables $z_{\alpha\beta}$. For Z_1, $Z_2 \in \mathbb{H}$, since Z'_2 is symmetric, we get

$$\begin{aligned} Z'_2 - Z'_1 &= (Z_2{}^tC + {}^tD)^{-1}(Z_2{}^tA + {}^tB) - (AZ_1 + B)(CZ_1 + D)^{-1} \\ &= \mu(M)(Z_2{}^tC + {}^tD)^{-1}(Z_2 - Z_1)(CZ_1 + D)^{-1}, \end{aligned}$$

where we have also used the relations (1.2). It follows that

$$\boldsymbol{D}Z' = \mu(M)(Z^tC + {}^tD)^{-1}\boldsymbol{D}Z(CZ+D)^{-1},$$

where $\boldsymbol{D}Z = (dz_{\alpha\beta})$ and $\boldsymbol{D}Z' = (dz'_{\gamma\delta})$ are the matrices of differentials of the variables $z_{\alpha\beta}$ and $z'_{\gamma\delta}$, respectively. Note that if $\rho(U)$ with $U \in \mathrm{GL}_n(\mathbb{C})$ is the transformation $(v_{\alpha\beta}) \mapsto U(v_{\alpha\beta})^tU$ of variables $v_{\alpha\beta} = v_{\beta\alpha}$ with $1 \le \alpha, \beta \le n$, then $\det\rho(U) = (\det U)^{n+1}$. This can be easily checked by ordering the variables $v_{\alpha\beta}$ lexicographically and replacing U by an upper triangular matrix of the form $W^{-1}UW$. Let $\boldsymbol{d}Z$ and $\boldsymbol{d}Z'$ be the columns with entries $dz_{\alpha\beta}$ $(1 \le \alpha, \beta \le n)$ and $dz'_{\gamma\delta}$ $(1 \le \gamma, \le \delta \le n)$ arranged in a fixed order. Then the above considerations imply the relation

$$\boldsymbol{d}Z' = \rho(\sqrt{\mu(M)}{}^t(CZ+D)^{-1})\boldsymbol{d}Z.$$

Taking $\boldsymbol{d}Z = \boldsymbol{d}X + i\boldsymbol{d}Y$, $\boldsymbol{d}Z' = \boldsymbol{d}X' + i\boldsymbol{d}Y'$, and $\rho(\sqrt{\mu(M)}{}^t(CZ+D)^{-1}) = R + iS$, we obtain that

$$\boldsymbol{d}X' = R\boldsymbol{d}X - S\boldsymbol{d}Y, \quad \boldsymbol{d}Y' = S\boldsymbol{d}X + R\boldsymbol{d}Y.$$

Thus, the Jacobian equals

$$\begin{aligned}\det\begin{pmatrix} R & -S \\ S & R\end{pmatrix} &= \det\left(\begin{pmatrix} 1_n & i1_n \\ 0 & 1_n\end{pmatrix}\begin{pmatrix} R & -S \\ S & R\end{pmatrix}\begin{pmatrix} 1_n & -i1_n \\ 0 & 1_n\end{pmatrix}\right)\\ &= \det\begin{pmatrix} R+iS & 0 \\ 0 & R-iS\end{pmatrix} = \mu(M)^{n(n+1)}|\det(CZ+D)|^{-2n-2}.\end{aligned}$$

□

By combining the above lemma and formula (1.10), we get the following result.

Proposition 1.6. *The element of volume on* $\mathbb{H}$ *given by*

$$d^*Z = \det Y^{-(n+1)}dZ \qquad (Z = X + iY \in \mathbb{H}), \tag{1.12}$$

where $dZ = dXdY$ *is the Euclidean element of volume* (1.11)*, is invariant under all transformations of the group* $\mathbb{G}$ *:*

$$d^*M\langle Z\rangle = d^*Z \qquad (M \in \mathbb{G}).$$

It is easy to see that two matrices M, M' of $\mathbb{G}$ have the same action (1.6) on an open subset of $\mathbb{H}$ if and only if $M' = \lambda M$ with λ in the set $\mathbb{R}^*$ of nonzero real numbers. It follows that the group of all transformations of the upper half-plane of the form (1.6) is isomorphic to the factor groups

$$\mathbb{G}/\{\mathbb{R}^*1_{2n}\} \simeq \mathbb{S}/\{\pm 1_{2n}\},$$

where

$$\mathbb{S} = \mathbb{S}^n = \mathrm{Sp}_n(\mathbb{R}) = \left\{M \in \mathbb{G}^n \;\middle|\; \mu(M) = 1\right\} \tag{1.13}$$

is the *(real) symplectic group of genus n*. We have already seen in the proof of Lemma 1.3 that each matrix $Z \in \mathbb{H}$ can be written in the form $Z = M\langle iE\rangle$ with a matrix $M \in \mathbb{S}$. Therefore, the upper half-plane can be identified with the homogeneous space of the symplectic group by the stabilizer $\mathbb{U}$ of the point iE in $\mathbb{S}$. More precisely, we have the following lemma.

Lemma 1.7. *The map $M \mapsto M\langle i1_n\rangle$ defines a one-to-one correspondence $\mathbb{S}/\mathbb{U} \leftrightarrow \mathbb{H}$, which is compatible with the actions of the group $\mathbb{S}$, where on the left side it acts by multiplication from the left; the stabilizer $\mathbb{U}$ has the form*

$$\mathbb{U} = \mathbb{U}^n = \left\{ \begin{pmatrix} A & B \\ -B & A \end{pmatrix} \in \mathbb{S}^n \right\};$$

the map $\left(\begin{smallmatrix} A & B \\ -B & A \end{smallmatrix}\right) \mapsto A + iB$ is an isomorphism of $\mathbb{U}$ onto the unitary group of order n; in particular, the group $\mathbb{U}$ is compact.

Exercise 1.8. Prove the lemma and preceding assertions.

Exercise 1.9. Show that the Cayley mapping

$$Z \mapsto W = (Z - i1_n)(Z + i1_n)^{-1} \qquad (Z \in \mathbb{H}),$$

is an analytic isomorphism of $\mathbb{H}$ onto the bounded domain

$$\left\{ W \in \mathbb{C}_n^n \,\middle|\, {}^tW = W, \quad \overline{W}W < E \right\},$$

where the inequality is understood in the sense of Hermitian matrices. Show that the inverse mapping is given by

$$W \mapsto Z = i(1_n + W)(1_n - W)^{-1}.$$

1.2 Fundamental Domains for the Modular Group

Modular Group. The *modular (symplectic) group* or the *Siegel modular group of genus n*, i.e., the group of all integral symplectic matrices of order $2n$ with unit multiplier,

$$\Gamma = \Gamma^n = \mathrm{Sp}_n(\mathbb{Z}) = \mathbb{S}^n \bigcap \mathbb{Z}_{2n}^{2n}, \qquad (1.14)$$

is clearly a discrete subgroup of the symplectic group $\mathbb{S}$. The same is true for each subgroup K of $\mathbb{S}$ *commensurable with the group* Γ, i.e., such that the intersection $K \cap \Gamma$ is of finite index both in K and Γ. Lemma 1.7 implies then that each such group K acts discretely on the upper half-plane.

Automorphic forms for subgroups K of the symplectic group, which we are going to consider in this chapter, are functions on the upper half-plane having certain analytic properties and satisfying functional equations connecting its values at points of each *K-orbit*

$$K\langle Z\rangle = \left\{ M\langle Z\rangle \,\middle|\, M \in K \right\} \qquad (Z \in \mathbb{H})$$

of K on $\mathbb{H}$, so that such a function is uniquely determined by its restriction to any subset of $\mathbb{H}$ that meets each K-orbit.

We recall that a closed subset $\mathbf{D}$ of a topological space $\mathbf{X}$ is called a *fundamental domain* for a discrete transformation group G acting on $\mathbf{X}$ if it meets each of the G-orbits $G(x) = \{g(x)|g \in G\}$ with $x \in \mathbf{X}$ and has no distinct inner points belonging to the same orbit. It follows from the definition that the decomposition

$$\mathbf{X} = \bigcup_{g \in G/G'} g(\mathbf{D}) \quad \text{with } G' = \left\{ g \in G \,\middle|\, g(x) = x, \quad \forall x \in X \right\} \tag{1.15}$$

holds, and its components pairwise have no common inner points. Fundamental domains do not necessarily exist.

Minkowski Reduction Domain. The construction of fundamental domains for the modular symplectic group is essentially based on the *Minkowski reduction theory of positive definite quadratic forms.* In matrix language, the problem of reduction of positive definite quadratic forms relative to unimodular equivalence is that of construction of a fundamental domain for the *unimodular group*

$$\Lambda = \Lambda^n = \mathrm{GL}_n(\mathbb{Z}) \tag{1.16}$$

acting on the cone

$$\mathbf{P} = \mathbf{P}_n = \left\{ Y \in \mathbb{R}^n_n \,\middle|\, {}^tY = Y, \quad Y > 0 \right\} \tag{1.17}$$

of real positive definite matrices of order n by

$$\Lambda \ni V : \quad Y \mapsto Y[V] = {}^tVYV.$$

The columns of a matrix $U \in \Lambda$ will be denoted by $\mathbf{u}_1, \ldots, \mathbf{u}_n$, so that $U = (\mathbf{u}_1, \ldots, \mathbf{u}_n)$. In order to choose a special representative $Y[U]$ of the *orbit*

$$\Lambda(Y) = \left\{ Y[V] \,\middle|\, V \in \Lambda \right\}$$

of a point $Y \in \mathbf{P}$, we determine the matrix U column by column with the help of some minimal conditions. Let

$$\Lambda_r = \Lambda^n_r = \left\{ (\mathbf{u}_1, \ldots, \mathbf{u}_r) \in \mathbb{Z}^n_r \,\middle|\, (\mathbf{u}_1, \ldots, \mathbf{u}_r, *, \ldots, *) \in \Lambda \right\}$$

be the set of all integral $n \times r-$matrices composed of the first r columns of matrices in Λ. For a given $Y \in \boldsymbol{P}$, we choose $\mathbf{u}_1 \in \Lambda_1$ such that the value $Y[\mathbf{u}_1] = {}^t\mathbf{u}_1 Y \mathbf{u}_1$ of the quadratic form with the matrix Y on the column $\mathbf{u}_1$ is minimal; this can be done, since the form is positive definite. Next, we determine $\mathbf{u}_2$ such that $(\mathbf{u}_1, \mathbf{u}_2) \in \Lambda_2$ and the value $Y[\mathbf{u}_2]$ is minimal. On replacing $\mathbf{u}_2$ by $-\mathbf{u}_2$ if necessary, one may assume that ${}^t\mathbf{u}_1 Y \mathbf{u}_2 \geq 0$. Proceeding in the same way, at the r–th step we choose $\mathbf{u}_r$ such that $(\mathbf{u}_1, \ldots, \mathbf{u}_r) \in \Lambda_r$, $Y[\boldsymbol{u}_r]$ is minimal, and ${}^t\boldsymbol{u}_{r-1} Y \boldsymbol{u}_r \geq 0$. Finally, when $r = n$, we get a unimodular matrix $U = (\mathbf{u}_1, \ldots, \mathbf{u}_n) \in \Lambda$ and a matrix $T = (t_{\alpha\beta}) = Y[U] \in \Lambda(Y)$, which is called *Minkowski reduced*, or just *reduced*.

Let us determine the conditions for a positive definite matrix to be reduced in terms of its entries. First of all, by induction on n based on the Euclidean algorithm, one can easily prove the following.

Lemma 1.10. *An integral n-column* $\boldsymbol{u}$ *belongs to* Λ_1 *if and only if its entries are coprime.*

Also, as an easy exercise on multiplication of block matrices, we get the following lemma.

Lemma 1.11. *Two matrices* U, U' *of* Λ^n *have the same first* r *columns if and only if*

$$U' = U \begin{pmatrix} 1_r & B \\ 0 & D \end{pmatrix} \qquad \text{with } D \in \Lambda^{n-r} \text{ and } B \in \mathbb{Z}_r^{n-r}.$$

Let $U = (\mathbf{u}_1, \ldots, \mathbf{u}_n) \in \Lambda^n$. By Lemma 1.11, the set of all r–th columns of all matrices $U' \in \Lambda^n$ with the first columns $\mathbf{u}_1, \ldots, \mathbf{u}_{r-1}$ coincides with the set of columns of the form $U\mathbf{v}$, where $\mathbf{v}$ is an integral n-column whose last $n-r+1$ entries $v_r, \ldots, v_n$ form the first column of a matrix $D \in \Lambda^{n-r+1}$. By Lemma 1.10, the last condition means that the numbers $v_r, \ldots, v_n$ are coprime. Thus, if $U = (\mathbf{u}_1, \ldots, \mathbf{u}_n) \in \Lambda^n$ and $1 \leq r \leq n$, then

$$\left\{\mathbf{u} \in \mathbb{Z}^n \;\middle|\; (\mathbf{u}_1, \ldots, \mathbf{u}_{r-1}, \mathbf{u}) \in \Lambda_r^n\right\} = U\mathbf{V}_{r,n}, \tag{1.18}$$

where

$$\mathbf{V}_{r,n} = \left\{\mathbf{v} = {}^t(v_1, \ldots, v_n) \in \mathbb{Z}^n \;\middle|\; \gcd(v_r, \ldots, v_n) = 1\right\}.$$

By the definition and (1.18), we conclude that a matrix $T = (t_{\alpha\beta}) = Y[U]$ is reduced if and only if it satisfies the conditions

$$Y[U\mathbf{v}] \geq Y[\mathbf{u}_r], \quad \text{for all } \mathbf{v} \in \mathbf{V}_{r,n} \text{ and } 1 \leq r \leq n,$$

and

$${}^t\mathbf{u}_{r-1} Y \mathbf{u}_r \geq 0, \quad \text{for } 1 < r \leq n,$$

where $U = (\mathbf{u}_1, \ldots, \mathbf{u}_n)$. Since $Y[U] = T = (t_{\alpha\beta})$, we have $Y[\mathbf{u}_r] = t_{rr}$ and ${}^t\mathbf{u}_{r-1} Y \mathbf{u}_r = t_{r-1,r}$. Hence, the above conditions mean exactly that T belongs to the set

$$\begin{aligned}\mathbf{M} = \mathbf{M}_n = \{T = (t_{\alpha\beta}) \in \mathbf{P}_n \;\big|\; & t_{rr} \leq T[\mathbf{v}] \text{ for all } \mathbf{v} \in \mathbf{V}_{r,n} \, (1 \leq r \leq n); \\ & t_{r-1,r} \geq 0 \, (1 < r \leq n)\},\end{aligned} \tag{1.19}$$

called the *Minkowski reduction domain*.

Theorem 1.12. *Every orbit $\Lambda(Y)$ of the group Λ on $\mathbf{P}$ contains at least one and not more than finitely many points of the Minkowski domain $\mathbf{M}$. If T, T' are two inner points of $\mathbf{M}$, and $T' = T[U]$ with $U \in \Lambda$, then $U = \pm 1_n$; in particular, no different inner points of $\mathbf{M}$ belong to the same orbit. In other words, $\mathbf{M}$ is a fundamental domain of Λ on $\mathbf{P}$.*

Proof. The above consideration shows that for a given $Y \in \mathbf{P}$, there exists $U \in \Lambda$ such that $Y[U] \in \mathbf{M}$, and every column of such U can be chosen in finitely many ways.

Let us set

$$\mathbf{M}' = \mathbf{M}'_n = \{T \in \mathbf{P}_n \mid t_{rr} < T[\mathbf{v}], \mathbf{v} \in \mathbf{V}_{r,n}, \mathbf{v} \neq \pm \mathbf{e}_r \, (1 \le r \le n);$$
$$t_{r-1,r} > 0 (1 < r \le n)\},$$

where $\mathbf{e}_1, \ldots, \mathbf{e}_n$ are the columns of the unit matrix 1_n. It is clear that $\mathbf{M}' \in \mathbf{M}$ and each inner point of $\mathbf{M}$ is contained in $\mathbf{M}'$. If $T = (t_{\alpha\beta})$ and $T' = (t'_{\alpha\beta})$ belong to $\mathbf{M}'$ and $T' = T[U]$ with $U = (\mathbf{u}_1, \ldots, \mathbf{u}_n) \in \Lambda$, then $\mathbf{u}_1 \in \mathbf{V}_{1,n}$, whence $t'_{11} = T[\mathbf{u}_1] \ge t_{11}$, and similarly, $t_{11} \ge t'_{11}$. It follows that $t_{11} = t'_{11} = T[\mathbf{u}_1]$, and so $\mathbf{u}_1 = \pm \mathbf{e}_1$. Then $\mathbf{u}_2 \in \mathbf{V}_{2,n}$, and in the same way we conclude that $\mathbf{u}_2 = \pm \mathbf{e}_2$. By repeating the same arguments, we see that $\mathbf{u}_r = \pm \mathbf{e}_r$ for all $r = 1, \ldots, n$. Now the conditions

$$t_{r-1,r} > 0, \quad t'_{r-1,r} = {}^t\mathbf{u}_{r-1} T \mathbf{u}_r > 0 \quad (1 < r \le n)$$

imply that $\mathbf{u}_r = \mathbf{e}_r$ or $\mathbf{u}_r = -\mathbf{e}_r$ for $r = 1, \ldots, n$, and $T = T'$. □

The entries of reduced matrices $T = (t_{\alpha\beta})$ satisfy some useful inequalities. First of all, since $t_{rr} \le T[\mathbf{e}_{r+1}] = t_{r+1,r+1}$, it follows that

$$t_{11} \le t_{22} \le \cdots \le t_{nn}. \tag{1.20}$$

Then by $t_{ll} \le T[\mathbf{e}_r \pm \mathbf{e}_l] = t_{rr} \pm 2t_{rl} + t_{ll}$, where $1 \le r < l \le n$, we obtain

$$|2t_{rl}| \le t_{rr} \qquad \text{if } r \neq l. \tag{1.21}$$

Finally, Minkowski have proved a deeper inequality for reduced matrices, which we cite without proof:

$$t_{11}t_{22} \cdots t_{nn} \le c_n \det T \qquad (T = (t_{\alpha\beta}) \in \mathbf{M}_n), \tag{1.22}$$

where c_n is a positive constant depending only on n. The Minkowski inequality implies that every $T = (t_{\alpha\beta}) \in \mathbf{M}_n$ satisfies the inequality

$$T \ge \frac{1}{n^{n-1} c_n} \operatorname{diag}(t_{11}, t_{22}, \ldots, t_{nn}). \tag{1.23}$$

Indeed, let $\rho_1, \ldots, \rho_n$ be the eigenvalues of the matrix

$$T' = T[\operatorname{diag}(t_{11}^{-1/2}, t_{22}^{-1/2}, \ldots, t_{nn}^{-1/2})],$$

then $\rho_1+\cdots+\rho_n=n$, and by (1.22),

$$\rho_1\cdots\rho_n=(t_{11}t_{22}\cdots t_{nn})^{-1}\det T\geq 1/c_n;$$

it follows that $\rho_\alpha\leq n$ and $\rho_\alpha\geq 1/n^{n-1}c_n$ for $\alpha=1,\dots,n$, which implies (1.23).

Exercise 1.13. Show that

$$\mathbf{M}_2=\left\{\begin{pmatrix}t_{11}&t_{12}\\t_{12}&t_{22}\end{pmatrix}\in\mathbf{P}_2\;\middle|\;0\leq 2t_{12}\leq t_{11}\leq t_{22}\right\}.$$

Show that in the inequalities (1.22) one can take $c_1=1$ and $c_2=4/3$, and the values are minimal.

[Hint: For minimality of c_2, consider $T=\left(\begin{smallmatrix}1&1/2\\1/2&1\end{smallmatrix}\right)$].

Exercise 1.14. Two binary quadratic forms $f(x,y)$ and $f'(x,y)$ are said to be equivalent if $f'(x,y)=f(\alpha x+\beta y,\gamma x+\delta y)$ with $\left(\begin{smallmatrix}\alpha&\beta\\\gamma&\delta\end{smallmatrix}\right)\in\Lambda^2$. Show that the number of classes of equivalent positive definite quadratic forms $f(x,y)=ax^2+bxy+cy^2$ with integral coefficients a, b, c and a fixed discriminant $d=b^2-4ac$ is finite.

Construction of Fundamental Domains. Let us return to the action of the modular group (1.14) on the upper half-plane. We consider orbits $\Gamma\langle Z\rangle$ of the modular group on $\mathbb{H}$. For $Z=X+iY\in\mathbb{H}=\mathbb{H}^n$, we shall call the positive real number $\det Y$ the *height* of the point Z and denote it by $h(Z)$. By (1.10), we have

$$h(M\langle Z\rangle)=|\det(CZ+D)|^{-2}h(Z)\qquad\left(Z\in\mathbb{H},M=\begin{pmatrix}A&B\\C&D\end{pmatrix}\in\Gamma\right).\tag{1.24}$$

Lemma 1.15. *Each orbit of the group Γ on $\mathbb{H}$ contains points Z of maximal height. These points can be characterized by the inequalities*

$$|\det(CZ+D)|\geq 1\qquad\textit{for every}\quad\begin{pmatrix}*&*\\C&D\end{pmatrix}\in\Gamma.$$

Proof. In view of (1.24) we have to show that $|\det(CZ+D)|$ takes a minimum on each orbit. Note that for any $M=\left(\begin{smallmatrix}A&B\\C&D\end{smallmatrix}\right)\in\Gamma$ and $V\in\Lambda=\Lambda^n$, the product

$$\begin{pmatrix}{}^tV^{-1}&0\\0&V\end{pmatrix}M=\begin{pmatrix}{}^tV^{-1}A&{}^tV^{-1}B\\VC&VD\end{pmatrix}$$

also belongs to Γ. It follows that if (C,D) is the "second row" of a matrix of Λ, then the same is true for (VC,VD). Replacing M in

$$M\langle X+iY\rangle=X'+iY'$$

by the above product does not change the value $|\det(CZ+B)|$ and replaces the matrix $(Y')^{-1}$ by the matrix ${}^tV(Y')^{-1}V$. Therefore, we may assume that the positive definite matrix $(Y')^{-1}$ is Minkowski reduced.

Let us denote by $\mathbf{c}_r$ and $\mathbf{d}_r$ $(r=1,\ldots,n)$ the columns of the matrices ${}^tC$ and $X\,{}^tC+{}^tD$, respectively, and by $t_1,\ldots,t_n$ the diagonal entries of $(Y')^{-1}$. Then, by (1.10), we can write

$$(Y')^{-1}=(CZ+D)Y^{-1}(\overline{Z}{}^tC+{}^tD)=(CX+D)Y^{-1}(X\,{}^tC+{}^tD)+CY\,{}^tC,$$

whence for $r=1,\ldots,n$, we get

$$t_r=Y^{-1}[\mathbf{d}_r]+Y[\mathbf{c}_r]\geq\begin{cases}Y[\mathbf{c}_r]\\ Y^{-1}[\mathbf{d}_r]\end{cases}. \tag{1.25}$$

If for some r, the columns $\mathbf{c}_r$ and $\mathbf{d}_r$ are both zero, then the $r-$th column of the matrix (CD) is also zero, which is impossible, since M is nonsingular. Since $Y>0$, the value $Y[\mathbf{c}_r]$ assumes a positive minimum when Z is fixed and $\mathbf{c}_r$ is an arbitrary nonzero integral column. On the other hand, if $\mathbf{c}_r=0$, then $\mathbf{d}_r$ is the $r-$th column of ${}^tD$ and so is a nonzero integral column. Since $Y^{-1}>0$, the value $Y^{-1}[\mathbf{d}_r]$ also assumes a positive minimum. It follows then from (1.25) that the numbers $u_1,\ldots,u_n$ have a positive lower bound independent of M. The relations (1.22) and (1.24) imply the inequality

$$t_1t_2\cdots t_n\leq c_n(\det Y')^{-1}=c_n(\det Y)^{-1}|\det(CZ+D)|^2.$$

If we now assume that a condition $|\det(CZ+D)|\leq h$ is satisfied for an arbitrarily large number h, then it implies upper bounds for $t_1,t_2,\ldots,t_n$. Then from (1.25) we obtain upper bounds for entries of the columns $\mathbf{c}_r$ and $\mathbf{d}_r$ and hence for the entries of the matrices C and D. Therefore, the condition is satisfied only for finitely many pairs (C,D) if Z is fixed and h is a given large number. This proves the lemma. □

Theorem 1.16. *Let $\mathbf{D}=\mathbf{D}_n$ be the subset of matrices $Z=X+iY\in\mathbb{H}_n$ satisfying the following conditions:*

(1) $|\det(CZ+D)|\geq 1$ *for every* $\begin{pmatrix} * & * \\ C & D\end{pmatrix}\in\Gamma=\Gamma^n$;
(2) $Y\in\mathbf{M}_n$, *where $\mathbf{M}_n$ is the Minkowski reduction domain* (1.19);
(3) $X\in\mathbf{X}_n=\left\{X=(x_{\alpha\beta})\in\mathbb{R}_n^n\;\middle|\;{}^tX=X,\quad |x_{\alpha\beta}|\leq 1/2\,(1\leq\alpha,\beta\leq n)\right\}$.

Then $\mathbf{D}$ meets each Λ-orbit on $\mathbb{H}$, and $Z'=M\langle Z\rangle$ for two inner points $Z,Z'\in\mathbf{D}$ with $M\in\Gamma$ if and only if $M=\pm E$, i.e., $\mathbf{D}$ is a fundamental domain of Γ on $\mathbb{H}$.

If $Z=X=iY\in\mathbf{D}$, then

$$Y\geq b_nE\quad\text{and}\quad\sigma(Y^{-1})\leq n/b_n, \tag{1.26}$$

where b_n is a positive constant depending only on n, and σ denotes the trace.

The volume of $\mathbf{D}$ with respect to the invariant element of volume (1.12) *is finite.*

Proof. Let us consider the orbit $\Gamma\langle Z''\rangle$ of a point $Z'' \in \mathbb{H}$. By Lemma 1.15, the orbit contains a point $Z' = X' + iY'$ of maximal height, and the point satisfies the condition (1). Every transformation of the form

$$Z' \mapsto \begin{pmatrix} {}^tV & SV^{-1} \\ 0 & V^{-1} \end{pmatrix} = {}^tVZ'V + S = X'[V] + S + iY'[V]$$

with $V \in \Lambda = \Lambda^n$ and an integral symmetric matrix S of order n corresponds to a matrix of Γ and does not change the height of Z'. By Theorem 1.12, there is a matrix $V \in \Lambda$ such that the matrix $Y = Y'[V]$ belongs to $\boldsymbol{M}_n$. Also, clearly, there is S such that $X'[V] + S \in \mathbf{X}_n$. Then

$$Z = X + iY \in \Gamma\langle Z'\rangle \bigcap \mathbf{D}.$$

Suppose now that $Z' = M\langle Z\rangle$ for two points $Z, Z' \in \mathbf{D}$ with $M = \left(\begin{smallmatrix} A & B \\ C & D \end{smallmatrix}\right) \in \Gamma$. Then $h(Z') = h(Z)$, by Lemma 1.15. It follows from (1.24) that $|\det(CZ + D)| = 1$. On the other hand, since $Z = M^{-1}\langle Z'\rangle$, we conclude that $|\det(-{}^tCZ' + {}^tA)| = 1$ (see (1.1)). If $C \neq 0$, then the equations are nontrivial, and so the points Z, Z' belong to the boundary of $\mathbf{D}$. If $C = 0$, then M has the form

$$M = \begin{pmatrix} {}^tV & SV^{-1} \\ 0 & V^{-1} \end{pmatrix} \qquad \text{with } V \in \Lambda \text{ and } S = {}^tS \in \mathbb{Z}_n^n.$$

So we have

$$Z' = X' + iY' = X[V] + S + iY[V], \qquad \text{where} \quad X + iY = Z,$$

and in particular, $Y' = Y[V]$. Since Y and Y' are both in $\mathbf{M}_n$, it follows from Theorem 1.12 that Y and Y' are boundary points of $\mathbf{M}_n$ or $V = \pm E$. In the last case we have $X' = X + S$, whence $S = 0$, unless X and X' belong to the boundary of $\mathbf{X}_n$. We conclude that $M = \pm 1_{2n}$, unless Z and Z' are boundary points of $\mathbf{D}$.

Let $Z = (z_{\alpha\beta}) = (x_{\alpha\beta} + iy_{\alpha\beta}) \in \mathbf{D}$. the inequality $|\det(CZ + D)| \geq 1$ for the pair

$$(C, D) = \left(\begin{pmatrix} 1 & 0 \\ 0 & 0 \end{pmatrix}, \begin{pmatrix} 0 & 0 \\ 0 & 1_{n-1} \end{pmatrix}\right)$$

implies the inequality $|z_{11}| = \sqrt{x_{11}^2 + y_{11}^2} \geq 1$. Since $|x_{11}| \leq 1/2$, it follows that $y_{11}^2 \geq 3/4$, i.e., $y_{11} \geq \sqrt{3}/2$. The last inequality and the inequalities (1.20) imply that $y_{\alpha\alpha} \geq \sqrt{3}/2$ for $\alpha = 1, \ldots, n$. The first inequality of (1.26) follows then from (1.23) with $b_n = \sqrt{3}/2n^{n-1}c_n$. The inequality implies that each eigenvalue of Y^{-1} is not greater than $1/b_n$, which proves the second inequality.

Finally, by (1.12), Theorem 1.16, and (1.26) we have

$$v(\mathbf{D}_n) = \int_{\mathbf{D}_n} (\det Y)^{-n-1} dX dY \leq \int_{Y \in \mathbf{M}_n, Y \geq b_n E} (\det Y)^{-n-1} dY,$$

which, by (1.20), (1.21), and (1.22), can be estimated as

$$\leq \int_{\substack{b_n \leq y_{11} \leq y_{22} \leq \cdots \leq y_{nn};\\ |2y_{\alpha\beta}| \leq y_{\alpha\alpha}\,(\alpha \neq \beta)}} (c_n^{-1} y_{11} y_{22} \cdots y_{nn})^{-n-1} dY$$

$$\leq \int_{y_{11}, y_{22}, \cdots, y_{nn} \geq b_n} (c_n^{-1} y_{11} y_{22} \cdots y_{nn})^{-n-1} \left(\prod_{\alpha=1}^{n} y_{\alpha\alpha}^{n-\alpha} \right) dy_{11} dy_{22} \cdots dy_{nn}$$

$$= c' \prod_{\alpha=1}^{n} \int_{c_n}^{\infty} y^{-\alpha-1} dy < \infty.$$

□

Exercise 1.17. Prove that $\mathbf{D}_1$ is the so-called modular triangle,

$$\mathbf{D}_1 = \left\{ z = x + iy \in \mathbb{H} \;\middle|\; |x| \leq \frac{1}{2},\ |z| = x^2 + y^2 \geq 1 \right\},$$

and one can take $b_1 = \frac{\sqrt{3}}{2}$. Draw the modular triangle.

Two binary quadratic forms

$$Q(x,y) = q_{11}x^2 + q_{12}xy + q_{22}y^2 \text{ and } Q'(x,y) = q'_{11}x^2 + q'_{12}xy + q'_{22}y^2$$

are said to be *properly equivalent* (over $\mathbb{Z}$) if

$$Q'(x,y) = Q(ax+by, cx+dy) \qquad \text{with} \qquad \begin{pmatrix} a & b \\ c & d \end{pmatrix} \in \Gamma^1 = \mathrm{SL}_2(\mathbb{Z}). \tag{1.27}$$

Exercise 1.18. Show that any real positive definite quadratic form Q is properly equivalent to a form Q' whose coefficients satisfy the inequalities $|q'_{12}| \leq q'_{11} \leq q'_{22}$. Show that the cone given by the inequalities in the space of the coefficients of the form contains no distinct inner points corresponding to properly equivalent forms.

[Hint: Let ω and ω' be the roots belonging to $\mathbb{H}$ of the equations $Q(x, 1) = 0$ and $Q'(x, 1) = 0$, respectively. Show that (1.27) is equivalent with the conditions

$$(q'_{12})^2 - 4q'_{11}q'_{22} = (q_{12})^2 - 4q_{11}q_{22} \quad \text{and} \quad \omega' = M^{-1}\langle\omega\rangle \quad \text{with} \begin{pmatrix} a & b \\ c & d \end{pmatrix} \in \Lambda;$$

then use Theorem 1.16 and Exercise 1.17.]

Theorem 1.19. *Each subgroup of the symplectic group* $\mathbb{S} = \mathbb{S}^n$ *of the form* $K_M = M^{-1}KM$, *where* K *is a subgroup of finite index in the modular group* $\Gamma = \Gamma^n$, *and* M *belongs to the general symplectic group* $\mathbb{G} = \mathbb{G}^n$, *has a fundamental domain* $\mathbf{D}(K_M)$ *on* $\mathbb{H} = \mathbb{H}^n$, *and one can take*

$$\mathbf{D}(K_M) = \bigcup_{\gamma \in \check{K} \backslash \Gamma} (M^{-1}\gamma)\langle \mathbf{D} \rangle, \tag{1.28}$$

where $\mathbf{D} = \mathbf{D}(\Gamma)$ *is a fundamental domain of* Γ, *and* γ *ranges over a system of representatives of different left cosets of* Γ *modulo the subgroup* $\check{K} = K \cup (-1_{2n})K$.

The invariant volume of the fundamental domain is finite.

Proof. First of all, the set (1.28) is closed, as a finite union of closed subsets. Next, if $(M^{-1}\gamma)\langle Z\rangle$ and $(M^{-1}\gamma')\langle Z'\rangle$ are two inner points of the set belonging to the same orbit of K_M, i.e., $(M^{-1}\gamma)\langle Z\rangle = (M^{-1}\delta M)\langle(M^{-1}\gamma')\langle Z'\rangle\rangle = (M^{-1}\delta\gamma')\langle Z'\rangle$ with $\delta \in K$, one can assume that Z and Z' are inner points of $\mathbf{D}$. Then, by (1.7), $\gamma\langle Z\rangle = (\delta\gamma')\langle Z'\rangle$, and so $Z = Z'$, and $\gamma = \pm\delta\gamma'$, by the definition of $\mathbf{D}$, which implies that $\gamma = \gamma'$ and $\delta = \pm 1_{2n}$. Finally, since $\mathbf{D}$ meets each Γ-orbit, we have

$$\mathbb{H} = \bigcup_{\delta\in\Gamma/\{\pm 1_{2n}\}} \delta\langle\mathbf{D}\rangle = \bigcup_{\delta\in\check{K}/\{\pm 1_{2n}\}} \bigcup_{\gamma\in\check{K}\backslash\Gamma} (\delta\gamma)\langle\mathbf{D}\rangle,$$

whence

$$\mathbb{H} = M^{-1}\langle\mathbb{H}\rangle = \bigcup_{\delta\in\check{K}/\{\pm 1_{2n}\}} \bigcup_{\gamma\in\check{K}\backslash\Gamma} ((M^{-1}\delta M)(M^{-1}\gamma))\langle\mathbf{D}\rangle.$$

This proves the decomposition (1.15), whence the set $\mathbf{D}(K_M)$ really meets each K_M-orbit. Finally, the set (1.28) is a finite union of images of the domain $\mathbf{D}$ under symplectic transformations and hence together with this domain has finite volume. □

1.3 Modular Forms

Definition of Modular Forms. Acting on the upper half-plane $\mathbb{H} = \mathbb{H}^n$, the general symplectic group $\mathbb{G} = \mathbb{G}^n$ operates also on complex-valued functions F on $\mathbb{H}$ by *Petersson operators of integral weights k*:

$$\mathbb{G} \ni M = \begin{pmatrix} A & B \\ C & D \end{pmatrix} : F \mapsto F|_k M = \mu(M)^{nk-\frac{n(n+1)}{2}} j(M,Z)^{-k} F(M\langle Z\rangle)$$

$$= \mu(M)^{nk-\frac{n(n+1)}{2}} \det(CZ+D)^{-k} F((AZ+B)(CZ+D)^{-1}), \tag{1.29}$$

where

$$j(M,Z) = \det J(M,Z) = \det(CZ+D).$$

Note that the factor $\mu(M)^{nk-\frac{n(n+1)}{2}}$ is introduced for aesthetic reasons, because it simplifies a number of forthcoming formulas. In fact, it can be replaced by an arbitrary quasicharacter of $\mathbb{G}$ or just omitted without affecting essential properties of the operators. By Lemma 1.3, the Petersson operators map holomorphic functions to holomorphic, moreover, it follows from the definition and relations (1.7), (1.9) that the operators satisfy the rules

$$F|_k rM = r^{n(k-n-1)} F|_k M \quad \text{for all } r \in \mathbb{R},\ r \neq 0 \tag{1.30}$$

and for $M, M' \in G^n$

$$\begin{aligned} F|_k MM' &= \mu(MM')^{nk-\frac{n(n+1)}{2}} j(MM', Z)^{-k} F((MM')\langle Z\rangle) \\ &= (\mu(M)\mu(M'))^{nk-\frac{n(n+1)}{2}} (j(M, M'\langle Z\rangle) j(M', Z))^{-k} F(M\langle M'\langle Z\rangle\rangle) \\ &= \mu(M')^{nk-\frac{n(n+1)}{2}} j(M', Z)^{-k} (F|_k M)(M'\langle Z\rangle) = (F|_k M)|_k M'. \quad (1.31) \end{aligned}$$

Let K be a subgroup of $\mathbb{G}$ commensurable with the modular group $\Gamma = \Gamma^n$, χ a *character* of K, that is, a multiplicative homomorphism of K into the nonzero complex numbers with kernel of finite index in K, and k an integer. A complex-valued function F on $\mathbb{H}$ is called a *(Siegel) modular form of weight k and character χ for the group K* if the following conditions are satisfied:

(i) F is a holomorphic function in $n(n+1)/2$ complex variables on $\mathbb{H}$;
(ii) for every matrix $M \in K$, the function F satisfies the functional equation

$$F|_k M = \chi(M) F, \tag{1.32}$$

where $|_k$ is the Petersson operator of weight k;
(iii) if $n = 1$, then every function $F|_k M$ with $M \in \Gamma^1$ is bounded on each subset of $\mathbb{H}^1$ of thc form $\mathbb{H}^1_\varepsilon - \{x + iy \in \mathbb{H}^1 | y \geq \varepsilon\}$ with $\varepsilon > 0$.

The set $\mathfrak{M}_k(K, \chi)$ of all modular forms of weight k and character χ for the group K is clearly a linear space over the field $\mathbb{C}$.

Exercise 1.20. Let $k > 2$ be an integer. Show that the Eisenstein series

$$E_k(z) = \sum_{\substack{c,d \in \mathbb{Z} \\ (c,d) \neq (0,0)}} \frac{1}{(cz+d)^k} \qquad (z \in \mathbb{H}^1)$$

converges absolutely and defines a modular form of weight k and the unit character for Γ^1.

[Hint: If $\mathbf{S}$ is a compact subset of $\mathbb{H}^1$, then $|\alpha z + \beta| \geq b(|\alpha| + |\beta|)$ for all $\alpha, \beta \in \mathbb{R}$ and $z \in \mathbf{S}$ with a positive constant b. Since there are only $4r$ pairs of integers (c, d) with $|c| + |d| = r$, it follows that $E_k(z)$ is dominated term by term on $\mathbf{S}$ by the series $b^{-k} \sum_{r=1}^{\infty} 4r/r^k$.]

For $q \in \mathbb{N}$, we shall denote by

$$\Gamma(q) = \Gamma^n(q) = \left\{ M \in \Gamma^n \,\middle|\, M \equiv 1_{2n} \pmod{q} \right\}$$

the *principal congruence subgroup of level q* of the modular group. Considering matrices of Γ modulo q, we get a homomorphism of the modular group into the finite group of symplectic matrices of order $2n$ over the residue class ring $\mathbb{Z}/q\mathbb{Z}$ with kernel $\Gamma(q)$. It follows that $\Gamma(q)$ is a normal subgroup of the modular group of finite index.

A subgroup K of the symplectic group (1.13) is called a *congruence subgroup* if it contains a principal congruence subgroup as a subgroup of finite index. A character of such K is said to be a *congruence character* if it is trivial on a principal congruence subgroup contained in K.

Proposition 1.21. *Let K be a congruence subgroup of $\mathbb{S}^n$, χ a congruence character of K, and M a matrix of $\mathbb{G}^n$ with rational entries then the group*

$$K_M = M^{-1}KM \tag{1.33}$$

is again a congruence subgroup of the symplectic group, the character

$$M^{-1}KM \ni M' \mapsto \chi_M(M') = \chi(MM'M^{-1}) \tag{1.34}$$

of this group is a congruence character, and the image $F|_kM$ of each modular form $F \in \mathfrak{M}_k(K,\chi)$ under the Petersson operator $|_kM$ is a modular form of weight k and character χ_M for the group K_M:

$$F|_kM \in \mathfrak{M}_k(K_M, \chi_M).$$

Proof. Let q be a positive integer such that $\Gamma^n(q) \subset K$ and χ is trivial on $\Gamma^n(q)$. It can be assumed that the matrix M is integral. Let $\mu(M) = q'$ then it is easy to see that the group $\Gamma^n(qq')$ is contained in K_M and the character χ_M is trivial on this group. It follows from definitions and Lemma 1.3 that the function $F' = F|_kM$ is holomorphic on $\mathbb{H}^n$. If $M' \in K_M$, then $MM'M^{-1} \in K$, and by (1.31), we have

$$\begin{aligned} F'|_kM' &= F|_kM|_kM' = F|_kMM'M^{-1}M = F|_kMM'M^{-1}|_kM \\ &= \chi(MM'M^{-1})F|_kM = \chi_M(M')F'. \end{aligned}$$

Finally, if $n=1$ and $V \in \Gamma^1$, we can write $F'|_kV = F|_kMV$ and present the matrix MV in the form $MV = WM_1$ with $W \in \Gamma^1 = \mathrm{SL}_2(\mathbb{Z})$ and an upper triangular matrix M_1. Then the function $F'|_kV = F|_kWM_1 = (F|_kW)|_kM_1$ together with $F|_kW$ is clearly bounded on each subset of $\mathbb{H}^1$ of the form $\mathbb{H}^1_\varepsilon$ with $\varepsilon > 0$. □

Exercise 1.22. Let $k > 2$ be an integer, $a, b \in \mathbb{Z}$, and $q \in \mathbb{N}$. Show that the Eisenstein series

$$E_k(z;(a,b)|q) = \sum_{\substack{c,d\in\mathbb{Z},\,(c,d)\neq(0,0),\\ (c,d)\equiv(a,b) \pmod{q}}} \frac{1}{(cz+d)^k} \qquad (z \in \mathbb{H}^1)$$

converges absolutely and defines a modular form of weight k and the unit character for $\Gamma^1(q)$ satisfying

$$E_k(z;(a,b)|q)|_kM = E_k(z;(a,b)M|q) \quad \text{for each } M \in \Gamma^1.$$

[Hint: Use Exercise 1.20.]

Fourier Expansion of Modular Forms.

Theorem 1.23. *Let K be a congruence subgroup of $\mathbb{S}^n$, and χ a congruence character of K. Then each modular form $F \in \mathfrak{M}_k(K,\chi)$ has an expansion of the form*

$$F(Z) = \sum_{A \in \mathbb{E}^n, A \geq 0} f(A) e^{\frac{\pi i}{q}\sigma(AZ)}, \tag{1.35}$$

with constant coefficients $f(A)$, where

$$\mathbb{E}^n = \left\{ A = (a_{\alpha\beta}) \in \mathbb{Z}_n^n \;\middle|\; {}^tA = A, \quad a_{11}, a_{22}, \dots, a_{nn} \in 2\mathbb{Z} \right\}$$

is the set of "even" matrices of order n, σ denotes the trace, and where $q = q(K,\chi)$ is a positive integer such that the group K contains a subgroup of the form

$$\mathbb{T}(q) = \mathbb{T}^n(q) = \left\{ \begin{pmatrix} 1_n & qB \\ 0 & 1_n \end{pmatrix} \;\middle|\; B = {}^tB \in \mathbb{Z}_n^n \right\}. \tag{1.36}$$

The series (1.35) converges absolutely on $\mathbb{H}$ and uniformly on each subset of $\mathbb{H}$ of the form

$$\mathbb{H}_\varepsilon = \mathbb{H}_\varepsilon^n = \left\{ X + iY \in \mathbb{H}^n \;\middle|\; Y \geq \varepsilon 1_{2n} \right\} \quad \textit{with} \quad \varepsilon > 0; \tag{1.37}$$

in particular, F is bounded on each of the subsets.

The coefficients $f(A)$ satisfy the relations

$$f({}^tVAV) = (\det V)^k \chi(M) e^{-\frac{\pi i}{q}\sigma(AVU)} f(A) \qquad (\forall A \in \mathbb{E}^n), \tag{1.38}$$

for every matrix M of the group K of the form

$$M = M(U,V) = \begin{pmatrix} V^{-1} & U \\ 0 & {}^tV \end{pmatrix}. \tag{1.39}$$

The expansion (1.35) is called the *Fourier expansion of F*, and the numbers $f(A)$ with $A \in \mathbb{E}^n$, $A \geq 0$ are the *Fourier coefficients of F*.

Proof. The functional equations (1.32) for matrices of the subgroup $\mathbb{T}(q) \subset K$ become

$$F(Z + qB) = F(Z) \qquad (Z = (z_{\alpha\beta}) \in \mathbb{H}, B = {}^tB \in \mathbb{Z}_n^n).$$

This means that F is periodic of period q in each of the variables $z_{\alpha\beta} = z_{\beta\alpha}$. Since F is also holomorphic, it can be expanded in a Fourier series of the form

$$F(Z) = \sum_{A'} f'(A';Y) \exp\left(\frac{2\pi i}{q} \sum_{1 \leq \alpha < \beta \leq n} a_{\alpha\beta} x_{\alpha\beta} \right),$$

where $Z = X + iY$ with $X = (x_{\alpha\beta})$, and $A' = (a_{\alpha\beta})$ ranges over the set of all upper triangular matrices of order n with integral entries $a_{\alpha\beta}$, and the series can be differentiated with respect to all of the variables. This expansion can be rewritten in the form

$$F(Z) = \sum_{A \in \mathbb{E}} f(A;Y) e^{\frac{\pi i}{q}\sigma(AZ)},$$

where $A = A' + {}^tA'$ runs through the set $\mathbb{E} = \mathbb{E}^n$ of all even matrices of order n, and where $f(A;Y) = f'(A';Y)\exp\left(-\frac{\pi}{q}\sigma(AY)\right)$. Since $F(Z)$ is holomorphic in each of the variables $z_{\alpha\beta}$, it satisfies the Cauchy–Riemann equations with respect to $z_{\alpha\beta}$, that is

$$\frac{\partial F}{\partial \bar{z}_{\alpha\beta}} = 0, \quad \text{where} \quad \frac{\partial}{\partial \bar{z}} = \frac{1}{2}\left(\frac{\partial}{\partial x} + i\frac{\partial}{\partial y}\right).$$

Since the last sum can be differentiated term by term, it follows that

$$\frac{\partial f(A,Y)}{\partial \bar{z}_{\alpha\beta}} = \frac{i}{2}\frac{\partial f(A,Y)}{\partial y_{\alpha\beta}} = 0, \quad \text{for } 1 \le \alpha \le \beta \le n.$$

Hence the coefficients $f(A,Y) = f(A)$ are independent of Y, and we get the expansion

$$F(Z) = \sum_{A \in \mathbb{E}} f(A) e^{\frac{\pi i}{q}\sigma(AZ)} \tag{1.40}$$

with constant coefficients.

The last expression can be considered as a Laurent expansion of the holomorphic function F in the variables $t_{\alpha\beta} = \exp(2\pi i z_{\alpha\beta}/q)$, and so it converges absolutely on $\mathbb{H}$.

The functional equations (1.32) for a matrix M of the form (1.39) give the relation

$$(\det V)^{-k} \sum_{A \in \mathbb{E}} f(A) e^{\frac{\pi i}{q}\sigma(AV^{-1}Z{}^tV^{-1} + AU{}^tV^{-1})} = \chi(M) \sum_{A \in \mathbb{E}} f(A) e^{\frac{\pi i}{q}\sigma(AZ)}.$$

On replacing A by ${}^tVAV$ on the left and comparing the coefficients, we get the relations (1.38).

In order to complete the proof, we have only to show that $f(A) = 0$, unless $A \ge 0$, and that the series converges uniformly on subsets (1.37).

If $n = 1$, the expansion (1.40) turns into the expansion

$$F(z) = \sum_{a \in 2\mathbb{Z}} f(a) e^{\frac{\pi i a z}{q}} = \sum_{a \in 2\mathbb{Z}} f(a) t^{a/2}, \qquad \left(t = e^{\frac{2\pi i z}{q}}\right), \tag{1.41}$$

which can be considered as a Laurent expansion of a function in t holomorphic for $|t| < 1$, except possibly at $t = 0$. Condition (iii) of the definition of modular forms for $n = 1$ implies that F is bounded in the circle and so is holomorphic at $t = 0$. Hence $f(a) = 0$ if $a < 0$, and the series converges uniformly on $\mathbb{H}^1_\varepsilon$ with $\varepsilon > 0$.

Now let $n \ge 2$. It was found by Max Koecher that in this case, an analogue of condition (iii) of the definition of modular forms is fulfilled automatically (*Koecher's effect*). We use the same arguments. Let Λ' be the group of all matrices $V \in \Lambda = \mathrm{SL}_n(\mathbb{Z})$ such that the matrix $M(0,V)$ of the form (1.39) with $U = 0$ belongs to the kernel of the character χ. Since χ is a congruence character, the group Λ' has finite index in Λ. By (1.38), we have $f({}^tVAV) = f(A)$, for all $A \in \mathbb{E}^n$ and $V \in \Lambda'$. Hence the expansion (1.40) can be rewritten in the form

$$F(Z) = \sum_{A \in \mathbb{E}/\Lambda'} f(A)\eta(Z, \{A\}),$$

where the sum is extended over a system of representatives for the *classes* $\{A\} = \{{}^tVAV \mid V \in \Lambda'\}$ of the set $\mathbb{E}$ modulo the equivalence $A \sim {}^tVAV$ with $V \in \Lambda'$, and where

$$\eta(Z, \{A\}) = \sum_{A' \in \{A\}} e^{\frac{\pi i}{q}\sigma(A'Z)}.$$

If $f(A) \neq 0$, then the series $f(A)\eta(Z, \{A\})$ converges absolutely for every $Z \in \mathbb{H}$, because it is a partial sum of an absolutely convergent series; in particular, the series

$$\eta(i1_n, \{A\}) = \sum_{A_i \in \{A\}} e^{-\frac{\pi}{q}\sigma(A_i)}$$

is convergent. Since the trace $\sigma(A_i)$ of every $A_i \in \{A\}$ is a rational integer, it follows that the class $\{A\}$ cannot contain more than a finite number of matrices A_i with $\sigma(A_i) < 0$. On the other hand, we shall show now that the trace $\sigma(A_i)$ assumes infinitely many negative values on the class $\{A\}$ of any integral symmetric matrix A of order $n \geq 2$, unless $A \geq 0$. If A does not satisfy $A \geq 0$, then there is an integral n-column $\mathbf{h}$ such that ${}^t\mathbf{h}A\mathbf{h} < 0$. For $r \in \mathbb{Z}$ we set

$$V_r = 1_n + rH \qquad \text{with} \quad H = (t_1\mathbf{h}, \ldots, t_n\mathbf{h}) \in \mathbb{Z}_n^n,$$

where $t_1, \ldots, t_n$ are some integers. Since the rank of the matrix rH is equal to 1 or 0, it follows that

$$\det V_r = 1 + \sigma(rH) = 1 + r(t_1h_1 + \cdots + t_nh_n),$$

where h_α are entries of $\mathbf{h}$. Since $n \geq 2$, there are integers $t_1, \ldots, t_n$ not all equal to 0 such that $t_1h_1 + \cdots + t_nh_n = 0$. Then we have

$$V_r \in \Lambda, \quad H^2 = ((t_1h_1 + \cdots + t_nh_n)h_\alpha t_\beta) = 0, \quad \text{and } V_r = V_1^r.$$

Since the index of Λ' in Λ is finite, it follows that the matrix $V_a = V_1^a$ belongs to Λ' for some $a \in \mathbb{N}$, and so do the matrices $V_{ab} = V_a^b$ for every integer b, hence the matrices ${}^tV_{ab}AV_{ab}$ belong to the class $\{A\}$, and

$$\begin{aligned}\sigma\left({}^tV_{ab}AV_{ab}\right) &= \sigma(A) + 2ab\sigma(AH) + a^2b^2\sigma({}^tVAV)\\ &= \sigma(A) + 2ab\sigma(AH) + a^2b^2({}^t\mathbf{h}A\mathbf{h})(t_1^2 + \cdots + t_n^2).\end{aligned}$$

The last expressions is a polynomial in b of degree 2 with negative leading coefficient. Hence it takes infinitely many negative values on $\mathbb{Z}$.

The above considerations show that a coefficient $f(A)$ in (1.30) equals zero, unless $A \geq 0$, which proves the expansion (1.35).

Finally, it is easy to see that the series (1.35) is majorized on each set (1.37) by a convergent series with nonnegative constant coefficients, and so converges there uniformly. □

Exercise 1.24. Show that if $k > 2$ is even, then the Fourier expansion of the Eisenstein series $E_k(z)$ defined in Exercise 1.20 has the form

$$E_k(z) = 2\sum_{n=1}^{\infty}\frac{1}{n^k} + 2\frac{(2\pi i)^k}{(k-1)!}\sum_{n=1}^{\infty}\left(\sum_{d|n} d^{k-1}\right)e^{2\pi i n z}.$$

[Hint: Show first that

$$\sum_{d\in\mathbb{Z}}\frac{1}{(z+d)^2} = \frac{\pi^2}{\sin^2 \pi z} = \frac{(2\pi)^2 t}{(1-t)^2} = (2\pi i)^2\sum_{d=1}^{\infty} d t^d,$$

where $t = e^{2\pi i z}$. Then differentiate both parts $k-2$ times.]

Cusp Forms. If K is a congruence subgroup, χ is a congruence character of K, and M is a matrix of the general symplectic group (1.4) with rational entries, then by Proposition 1.21 and Theorem 1.23, the function $F|_k M$ has a Fourier expansion of the form

$$(F|_k M)(Z) = \sum_{A\in\mathbb{E}^n, A\geq 0} f_M(A) e^{\frac{\pi i}{q}\sigma(AZ)} \tag{1.42}$$

with a positive integer q depending on K, χ, and M, which converges absolutely on $\mathbb{H}$ and uniformly on the subsets (1.37). The modular form F is called a *cusp form* if the coefficients $f_M(A)$ of the decomposition (1.42) satisfy the conditions

$$f_M(A) = 0 \quad \text{for all } M \in \mathbb{G}(\mathbb{Q}) \quad \text{and } A \in \mathbb{E} \quad \text{with } \det A = 0, \tag{1.43}$$

where

$$\mathbb{G}(\mathbb{Q}) = \mathbb{G}^n(\mathbb{Q}) = \mathbb{G}^n \cap \mathbb{Q}^{2n}_{2n}.$$

The subspace of cusp forms of $\mathfrak{M}_k(K, \chi)$ will be denoted by $\mathfrak{N}_k(K, \chi)$.

Proposition 1.25. *Let K be a congruence subgroup of $\mathbb{S} = \mathbb{S}^n$ and χ a congruence character of K. Then for each cusp form $F \in \mathfrak{N}_k(K, \chi)$ and each matrix $M \in \mathbb{G}^n(\mathbb{Q})$, the modular form $F|_k M \in \mathfrak{M}_k(K_M, \chi_M)$, where $K_M = M^{-1}KM$ and χ_M is the character* (1.34), *is a cusp form with a Fourier expansion of the form*

$$(F|_k M)(Z) = \sum_{A\in\mathbb{E}^n, A>0} f_M(A) e^{\frac{\pi i}{q}\sigma(AZ)}, \tag{1.44}$$

where q is a positive integer. If $k \geq 0$, then the form $F|_k M$ satisfies

$$|(F|_k M)(X+iY)| \leq c(\det Y)^{k/2} \qquad (X+iY \in \mathbb{H}), \tag{1.45}$$

and its Fourier coefficients satisfy

$$|f_M(A)| \le c'(\det A)^{k/2} \qquad \text{for all } A \in \mathbb{E} \quad \text{with } A > 0, \tag{1.46}$$

where c and c' are constants depending only on F and M.

First we shall prove the following simple lemma.

Lemma 1.26. *Let*

$$\phi(Y) = \sum_{A \in \mathbb{E}^n, A>0} \varphi(A) e^{-\eta\sigma(AY)},$$

where Y belongs to the cone $\mathbf{P} = \mathbf{P}_n$ of positive definite matrices of order n, and $\eta > 0$, be a series with nonnegative coefficients $\varphi(A)$ convergent for all $Y \in \mathbf{P}$. Then for every Minkowski reduced matrix Y satisfying $Y \ge d \cdot 1_n$ with $d > 0$, the estimate

$$\phi(Y) \le d' e^{-d''\sigma(Y)}$$

holds with positive constants d' and d''.

Proof of the lemma. If $Y = (y_{\alpha\beta}) \in \boldsymbol{P}$ and $A = (a_{\alpha\beta}) \in \mathbb{E}$ satisfies $A > 0$, then by (1.23),

$$\sigma(AY) \ge b\sigma(A\,\mathrm{diag}(y_{11}, y_{22}, \ldots, y_{nn})) = b\sum_{\alpha=1}^{n} a_{\alpha\alpha} y_{\alpha\alpha} \ge 2b\sigma(Y),$$

where b is a positive constant depending only on n. On the other hand, if $Y > dE$, then for $A = (a_{\alpha\beta}) \in \mathbb{E}$ with $A > 0$, we obtain $\sigma(AY) \ge d\sigma(A)$. (Note that we have used twice the obvious inequality $\sigma(AR) \ge \sigma(BR)$ valid if matrices $A - B$ and R are positive semidefinite.) It follows then from the above inequalities that

$$\sigma(AY) \ge b\sigma(Y) + \frac{1}{2}d\sigma(A),$$

hence

$$\phi(Y) \le \sum_{A \in \mathbb{E}, A>0} \varphi(A) e^{-\eta(b\sigma(Y) + \frac{1}{2}d\sigma(A))} = e^{-\eta b\sigma(Y)} \phi\left(\frac{1}{2}dE\right),$$

which proves the estimate. □

Proof of Proposition 1.25. The Fourier expansion (1.44) and the inclusion $F|_k M \in \mathfrak{N}_k(K_M, \chi_M)$ follow from Proposition 1.21 and the definition of cusp forms. By Proposition 1.21 there exists a level $q \in \mathbb{N}$ such that the principal congruence subgroup $\Gamma(q)$ is contained both in K_M and the kernel of character χ. Let us consider the function

$$G = \sum_{\gamma_i \in \Gamma(q)\backslash\Gamma} |F'|_k \gamma_i|, \quad \text{where } F' = F|_k M. \tag{1.47}$$

It follows from (1.31) and (1.32) that G is independent of the choice of representatives in the cosets $\Gamma(q)\backslash\Gamma$ and satisfies

$$G|_k\gamma = \sum_{\gamma_i\in\Gamma(q)\backslash\Gamma} |F'|_k\gamma_i\gamma| = G \qquad \text{for all } \gamma\in\Gamma, \tag{1.48}$$

because $\gamma_i\gamma$ runs through a system of representatives for $\Gamma(q)\backslash\Gamma$ when γ_i does. Hence by (1.10), we see that the function

$$H(Z) = H(X+iY) = (\det Y)^{k/2}G(Z) = (\det Y)^{k/2} \sum_{\gamma_i\in\Gamma(q)\backslash\Gamma} |F'|_k\gamma_i| \tag{1.49}$$

satisfies

$$H(\gamma\langle Z\rangle) = H(Z) \qquad \text{for all } Z\in\mathbb{H} \quad \text{and } \gamma\in\Gamma. \tag{1.50}$$

Since by Theorem 1.21, each of the functions $F'|_k\gamma_i = F|_kM\gamma_i$ is bounded on the sets $\mathbb{H}_\varepsilon$ with $\varepsilon > 0$, it follows from (1.26) that the functions are bounded on the fundamental domain $\mathbf{D}$ of Γ, defined in Theorem 1.16. Furthermore, each of the these functions satisfies

$$|F'|_k\gamma_i| \le \sum_{A\in\mathbb{E},A>0} |f_{M\gamma_i}(A)|e^{-\frac{\pi}{q}\sigma(AY)},$$

and the last series converges on the cone $\mathbf{P} = \mathbf{P}_n$. If $Z\in\mathbf{D}$, then Y is Minkowski reduced and satisfies $Y \ge b_n 1_n$; hence by Lemma 1.24, each of the series is dominated by a function of the form $d'\exp(-d''\sigma(Y))$ with constants d' and d'' depending on F and M. Therefore, since $k \ge 0$, we obtain

$$\begin{aligned} H(X+iY) &\le d'[\Gamma:\Gamma(q)](\det Y)^{k/2}\exp(-d''\sigma(Y)) \\ &\le \delta \prod_{\alpha=1}^{n} y_{\alpha\alpha}^{k/2}\exp(-d''y_{\alpha\alpha}), \end{aligned}$$

where $y_{\alpha\alpha}$ are the diagonal entries of Y, and we have also used the following consequence of the Hadamard's determinant theorem:

$$\det Y \le y_{11}y_{22}\cdots y_{nn} \qquad (Y\in\mathbf{P}_n). \tag{1.51}$$

Since the last expression is clearly bounded on $\mathbf{P}$, it follows that H is bounded on $\mathbf{D}$. Thus by (1.50), H is bounded on $\mathbb{H}$, which implies the estimate (1.45). The estimate for Fourier coefficients follows from (1.45). □

One can prove that the coefficients of the Fourier expansion (1.35) of an arbitrary modular form of nonnegative weight k and congruence character for a congruence subgroup of $\mathbb{S}^n$ satisfy

$$|f(A)| \le c(\det A)^k \qquad (A\in\mathbb{E}^n, A\ge 0), \tag{1.52}$$

where c depends only on the form.

The Siegel Operator. The general philosophy of modular forms emanates from the idea that consideration of arbitrary modular forms can usually be reduced to the case of cusp forms and the cases of modular forms of smaller genera. The reduction is ensured by the so-called Siegel operator and its iterations. Let $F = F(Z)$ be a modular form of weight k for a congruence subgroup K of the symplectic group $\mathbb{S} = \mathbb{S}^n$ and a congruence character χ of K. Since the Fourier series (1.25) for F converges uniformly on subsets of $\mathbb{H}^n$ of the form (1.27), the limit

$$(F|\Phi)(Z') = \lim_{y \to +\infty} F\left(\begin{pmatrix} Z' & 0 \\ 0 & iy \end{pmatrix}\right) = \sum_{A \in \mathbb{E}^n, A \geq O} f(A) \lim_{y \to +\infty} e^{\frac{\pi i}{q}\sigma(AZ'_y)}, \tag{1.53}$$

where $Z'_y = \begin{pmatrix} Z' & 0 \\ 0 & iy \end{pmatrix}$, exists for every $Z' \in \mathbb{H}^{n-1}$. If $A = \begin{pmatrix} A' & * \\ * & a_{nn} \end{pmatrix}$, then $\sigma(AZ'_y) = \sigma(A'Z') + iya_{nn}$, hence

$$\lim_{y \to +\infty} e^{\frac{\pi i}{q}\sigma(AZ'_y)} = \lim_{y \to +\infty} e^{-\frac{\pi}{q}ya_{nn}} e^{\frac{\pi i}{q}\sigma(A'Z')} = \begin{cases} e^{\frac{\pi i}{q}\sigma(A'Z')}, & \text{if } a_{nn} = 0, \\ 0, & \text{if } a_{nn} > 0. \end{cases}$$

Since $A \geq 0$, the equality $a_{nn} = 0$ implies that A has the form $\begin{pmatrix} A' & 0 \\ 0 & 0 \end{pmatrix}$. Thus we have

$$(F|\Phi)(Z') = \sum_{A' \in \mathbb{E}^{(n-1)}, A' \geq 0} f\left(\begin{pmatrix} A' & 0 \\ 0 & 0 \end{pmatrix}\right) e^{\frac{\pi i}{q}\sigma(A'Z')}, \tag{1.54}$$

for all $Z' \in \mathbb{H}^{(n-1)}$. The last series is a partial series for the Fourier expansion of F, and so it converges absolutely on $\mathbb{H}^{(n-1)}$ and uniformly on $\mathbb{H}_\varepsilon^{(n-1)}$ with $\varepsilon > 0$. If $n = 1$, we set

$$F|\Phi = \lim_{y \to +\infty} F(iy) \in \mathbb{C}. \tag{1.55}$$

As above, the limit exists and is equal to the constant term of the Fourier expansion of F. The linear operator

$$\Phi : F \mapsto F|\Phi \qquad (F \in \mathfrak{M}_k(K, \chi)) \tag{1.56}$$

is called the *Siegel operator.*

In order to show that the function $F|\Phi$ is a modular form on $\mathbb{H}^{(n-1)}$, we shall need new notation. Let $n > 1$. For a matrix $M' = \begin{pmatrix} A' & B' \\ C' & D' \end{pmatrix}$ with square blocks A', B', C', and D' of order $n-1$, we set

$$\overrightarrow{M'} = \begin{pmatrix} A & B \\ C & D \end{pmatrix} \qquad \text{and} \qquad \phi(\overrightarrow{M'}) = \begin{pmatrix} A' & B' \\ C' & D' \end{pmatrix} = M', \tag{1.57}$$

where $A = \begin{pmatrix} A' & 0 \\ 0 & 1 \end{pmatrix}$, $B = \begin{pmatrix} B' & 0 \\ 0 & 0 \end{pmatrix}$, $C = \begin{pmatrix} C' & 0 \\ 0 & 0 \end{pmatrix}$, $D = \begin{pmatrix} D' & 0 \\ 0 & 1 \end{pmatrix}$. If S is a subgroup of $\mathbb{G}^n$, we denote by $\overleftarrow{S}$ the set of all matrices $M' = \begin{pmatrix} A' & B' \\ C' & D' \end{pmatrix}$ such that the matrix $\overrightarrow{M'}$ belongs to S:

$$\overleftarrow{S} = \left\{ M' = \begin{pmatrix} A' & B' \\ C' & D' \end{pmatrix} \in \mathbb{R}^{2n-2}_{2n-2} \,\middle|\, \overrightarrow{M'} \in S \right\}.$$

Then it is clear that $\overleftarrow{S}$ is a subgroup of $\mathbb{S}^{(n-1)}$, and the map ψ given by

$$\overleftarrow{S} \ni M' \mapsto \psi(M') = \overrightarrow{M'} \in S \tag{1.58}$$

is a homomorphic embedding of the group $\overleftarrow{S}$ into S.

Lemma 1.27. *Let K be a congruence subgroup of $\mathbb{S}^n$ with $n > 1$, and χ a congruence character of K. Then the group $\overleftarrow{K}$ is a congruence subgroup of $\mathbb{G}^{(n-1)}$, and the map $\overleftarrow{\chi}$ given by*

$$\overleftarrow{K} \ni M' \mapsto \overleftarrow{\chi}(M') = \chi(\overrightarrow{M'}) \tag{1.59}$$

is a congruence character of the group $\overleftarrow{K}$.

Proof. By the assumptions, K contains a principal congruence subgroup $\Gamma(q) = \Gamma^n(q)$ of finite index belonging to kernel of χ. Then the group $\overleftarrow{\Gamma(q)}$ is clearly a principal congruence subgroup of $\overleftarrow{K}$ belonging to the kernel of the character $\overleftarrow{\chi}$. It remains to prove that $\overleftarrow{\Gamma(q)}$ has finite index in $\overleftarrow{K}$. Indeed, if $\overleftarrow{\Gamma(q)}M'_\alpha \neq \overleftarrow{\Gamma(q)}M'_\beta$, where M'_α, M'_β belong to $\overleftarrow{K}$, then $\Gamma(q)\overrightarrow{M'_\alpha} \neq \Gamma(q)\overrightarrow{M'_\beta}$, since otherwise, we would have $\overrightarrow{M'_\alpha}(\overrightarrow{M'_\beta})^{-1} \in \Gamma(q)$, and so $M'_\alpha(M'_\beta)^{-1} \in \overleftarrow{\Gamma}(q)$. □

Now we can prove our next result.

Theorem 1.28. *Let K be a congruence subgroup of $\mathbb{S}^n$, and χ a congruence character of K. Then the Siegel operator Φ maps the space $\mathfrak{M}_k(K,\chi)$ into the space $\mathfrak{M}_k(\overleftarrow{K},\overleftarrow{\chi})$:*

$$\Phi : \mathfrak{M}_k(K,\chi) \mapsto \mathfrak{M}_k(\overleftarrow{K},\overleftarrow{\chi}), \tag{1.60}$$

where for $n = 1$, we set $\mathfrak{M}_k(\overleftarrow{K},\overleftarrow{\chi}) = \mathbb{C}$.

Proof. One can assume that $n > 1$. Let $F \in \mathfrak{M}_k(K,\chi)$, $Z' \in \mathbb{H}^{n-1}$, $M' = \begin{pmatrix} A' & B' \\ C' & D' \end{pmatrix} \in \overleftarrow{\mathbb{S}}$, $Z'_\lambda = \begin{pmatrix} Z' & 0 \\ 0 & i\lambda \end{pmatrix}$, and $M = \begin{pmatrix} A & B \\ C & D \end{pmatrix} = \psi(M') \in \mathbb{S}$, where ψ is the embedding (1.58). Then we have

$$M\langle Z'_\lambda \rangle = \begin{pmatrix} A'Z' + B' & 0 \\ 0 & i\lambda \end{pmatrix} \begin{pmatrix} C'Z' + D' & 0 \\ 0 & 1 \end{pmatrix}^{-1} = M'\langle Z' \rangle_\lambda,$$

$$\det(CZ'_\lambda + D) = \det(C'Z' + D'), \quad \text{and} \quad \chi(M) = \overleftarrow{\chi}(M').$$

It follows that

$$\begin{aligned}(F|\Phi|_k M')(Z') &= \det(C'Z' + D')^{-k} \lim_{\lambda \to +\infty} F(M'\langle Z' \rangle_\lambda) \\ &= \lim_{\lambda \to +\infty} \det(CZ_\lambda + D)^{-k} F(M\langle Z'_\lambda \rangle) = (F|_k M|\Phi)(Z'). \end{aligned} \tag{1.61}$$

In particular, if $M' \in \overleftarrow{K}$, then we have

$$F|\Phi|_k M' = \chi(M) F\Phi = \overleftarrow{\chi}(M') F|\Phi.$$

Furthermore, it follows from (1.54) and (1.61) that the function $(F|\Phi|_k M')(Z')$ is holomorphic on $\mathbb{H}^{n-1}$, and is bounded on each subset $\mathbb{H}^{n-1}_\varepsilon$ with $\varepsilon > 0$. □

The next lemma gives useful characterizations of cusp forms in terms of Siegel operators.

Lemma 1.29. *Let K be a congruence subgroup of $\mathbb{S}^n$ and χ a congruence character of K and let $F \in \mathfrak{M}_k(K,\chi)$. Then the following three conditions are equivalent:*

(1) *The function F is a cusp form;*
(2) *F satisfies*

$$(F|_k M)|\Phi = 0 \qquad \text{for all } M \in \mathbb{G}^n(\mathbb{Q}) = \mathbb{G}^n \bigcap \mathbb{Q}^{2n}_{2n},$$

where Φ is the Siegel operator;
(3) *F satisfies*

$$(F|_k \gamma)|\Phi = 0 \qquad \text{for all } \gamma \in \Gamma^n.$$

Proof. It follows from definition of cusp forms and formula (1.54) that condition 1 implies condition 2. Condition 3 is a particular case of condition 2. Hence it remains to prove that each function $F \in \mathfrak{M}_k(K,\chi)$ satisfying condition 3 is a cusp form. It easily follows by an induction based on the Euclidean algorithm that each matrix $M \in \mathbb{G}^n(\mathbb{Q})$ can be written in the form $M = \gamma M_1$, where $\gamma \in \Gamma^n$ and M_1 has the form $M_1 = \begin{pmatrix} A_1 & B_1 \\ 0 & D_1 \end{pmatrix}$ with the zero $n \times n$-block $0 = 0_n$ (see Proposition 3.35(1) below). Since $F|_k M = F|_k \gamma M_1 = F|_k \gamma|_k M_1$, the Fourier expansion (1.42) of the function $F|_k M$ can be rewritten in the form

$$\sum_{A \in \mathbb{E}^n, A \geq 0} f_M(A) e^{\frac{\pi i}{q}\sigma(AZ)} = (F|_k M)(Z) = (\det D_1)^{-k} (F|_k \gamma)(A_1 Z D_1^{-1} + B_1 D_1^{-1})$$

$$= (\det D_1)^{-k} \sum_{A' \in \mathbb{E}^n, A' \geq 0} f_\gamma(A') e^{\frac{\pi i}{q_1}\sigma(A' B_1 D_1^{-1})} e^{\frac{\pi i}{q_1}\sigma(D_1^{-1} A' A_1 Z)},$$

where $f_\gamma(A')$ are the coefficients of the Fourier expansion (1.42) of the function $F|_k \gamma$. Hence if $f_\gamma(A') = 0$ for all $A' \in \mathbb{E}^n$ with $\det A' = 0$, then $f_M(A) = 0$ for all $A \in \mathbb{E}^n$ with $\det A = 0$. □

Spaces of Modular Forms. We can now prove the following important theorem.

Theorem 1.30. *Let K be a congruence subgroup of $\mathbb{S}^n$, χ a congruence character of K, and k a nonnegative integer. Then the space $\mathfrak{M}_k(K,\chi)$ of modular forms of weight k and character χ for the group K is finite-dimensional over the field $\mathbb{C}$.*

The proof of the theorem is based on the following key lemma.

Lemma 1.31. *Let*

$$F(Z) = \sum_{A\in\mathbb{E},A>0} f(A)e^{\pi i\sigma(AZ)}$$

be a cusp form of a nonnegative integral weight k and the unit character χ for the full modular group $\Gamma = \Gamma^n$. Suppose that the Fourier coefficients satisfy the conditions

$$f(A) = 0 \qquad \text{if} \quad \sigma(A) \leq \frac{kn}{2\pi b_n}, \tag{1.62}$$

where b_n is a constant satisfying the inequalities (1.26) *of Theorem* 1.16. *Then the form F is identically equal to zero.*

Proof of the lemma. By Lemma 1.4 and the definition of modular forms, the function

$$H(Z) = H(X+iY) = (\det Y)^{k/2}|F(Z)|$$

satisfies $H(\gamma\langle Z\rangle) = H(Z)$ for all $\gamma \in \Gamma$ and, as we have seen in the proof of Proposition 1.25, is bounded on the fundamental domain $\mathbf{D}_n$ of Γ defined in Theorem 1.16. Moreover, it follows from the estimates (1.51) and (1.52) that $H(X+iY) \to 0$, with $\det Y \to +\infty$ remaining in $\mathbf{D}_n$. It follows from Theorem 1.16 that any subset of $\mathbf{D}_n$ of the form $\{X+iY \in \mathbf{D}_n | \det Y \leq c\}$ with $c > 0$ is bounded and closed, and therefore is compact. Hence the function $H(Z)$ attains its maximum μ at some point $Z_0 = X_0 + iY_0$ of $\mathbf{D}_n$. Since H is Γ-invariant, we conclude that μ is the maximum of H on $\mathbb{H}$, that is, $H(Z) \leq H(Z_0) = \mu$ for all $Z \in \mathbb{H}$. Let us set $Z_t = Z_0 + tE$, where $t = u+iv$ is a complex parameter, and consider the function

$$\begin{aligned} h(t) &= F(Z_t)e^{-\pi i\lambda\sigma(Z_t)} \\ &= \sum_{A\in\mathbb{E},A>0} f(A)e^{\pi i(\sigma(AZ_0)+t\sigma(A))-\lambda\sigma(Z_0+tE))} \\ &= \sum_{A\in\mathbb{E},A>0} f(A)e^{\pi i(\sigma(AZ_0-\lambda Z_0))}e^{\pi it(\sigma(A)-\lambda n)} = h'(w), \end{aligned}$$

where $w = e^{\pi it}$ and λ satisfies $\lambda n = 1 + [kn/2\pi b_n]$, with $[\alpha]$ denoting the greatest integer not exceeding α. By the assumption of the lemma, we have $f(A) = 0$ if $\sigma(A) - \lambda n < 0$, and so the expansion of the function $h'(w)$ does not contain negative powers of w. If $\varepsilon > 0$ is so small that $Z_t \in \mathbb{H}$ for $v \geq -\varepsilon$, then the expansion converges absolutely and uniformly on the half-plane $v \geq -\varepsilon$, and so the function $h'(w)$ is holomorphic in the disk $|w| \leq e^{\pi\varepsilon} = \tau$. Since $\tau > 1$, it follows, by the maximum-modulus principle, that there is a point $w_0 = e^{\pi it_0}$ satisfying $|w_0| = \tau$ and $h'(1) \leq h'(w_0)$. Coming back to the function h, we can rewrite the last inequality in the form

$$|F(Z_0)e^{\pi\lambda\sigma(Y_0)}| \leq |F_{t_0}|e^{\pi\lambda\sigma(Y_0)}e^{\pi\lambda nv_0},$$

where $t_0 = u_0 + v_0$, hence

$$(\det Y_0)^{-k/2}H(Z_0) \leq (\det Y_{t_0})^{-k/2}H(Z_{t_0})e^{\pi\lambda nv_0}.$$

Since $H(Z_0)=\mu$ and $H(Z_{t_0})\le\mu$, the last inequality implies the inequality

$$\mu\le\mu(\det Y_0)^{k/2}(\det Y_{t_0})^{-k/2}e^{\pi\lambda n v_0}=\mu\psi(v_0),$$

where $\psi(v)=\det(E+vY_0^{-1})^{-k/2}e^{\pi\lambda n v}$. We have $\psi(0)=1$. Let us show that the derivative of ψ is positive at $v=0$. We can write

$$\psi(v)=e^{\pi\lambda n v}\prod_{j=1}^{n}(1+v\lambda_j)^{-k/2},$$

where $\lambda_1,\dots,\lambda_n$ are the eigenvalues of Y_0^{-1}, hence the value of the derivative at $v=0$ is

$$\begin{aligned}\pi\lambda n-\frac{k}{2}(\lambda_1+\dots+\lambda_n)&=\pi\lambda n-\frac{k}{2}\sigma(Y_0^{-1})\\ &\ge\pi\lambda n-\frac{kn}{2b_n}=\pi\left(1+\left[\frac{kn}{2\pi b_n}\right]\right)-\frac{kn}{2b_n}>0,\end{aligned}$$

by (1.26), since $X_0+iY_0\in\boldsymbol{D}_n$. It follows that for small $\varepsilon>0$, we have $\psi(v_0)=\psi(-\varepsilon)<1$ (we recall that $e^{\pi\varepsilon}=\tau=|e^{\pi i(u_0+iv_0)}|=e^{-\pi v_0}$). Then the above inequality shows that $\mu=0$, and so F is identically equal to zero. □

Proof of Theorem 1.30. Note, first of all, that if the character χ is trivial on a principal congruence subgroup $\Gamma(q)=\Gamma^n(q)$ contained in the group K, then the space $\mathfrak{M}_k(K,\chi)$ is contained in the space $\mathfrak{M}=\mathfrak{M}_k(\Gamma(q))$ of modular forms of weight k and the unit character for the group $\Gamma(q)$. Therefore, it will be sufficient to prove that each of these spaces is finite-dimensional. We recall that $\Gamma(q)$ is a normal subgroup of finite index $\nu=[\Gamma:\Gamma(q)]$ in Γ. Let $\gamma_1,\dots,\gamma_\nu$ be a system of representatives for cosets of Γ modulo $\Gamma(q)$. For a function $F\in\mathfrak{M}$, let us consider the functions

$$F_1=F|_k\gamma_1,\dots,F_\nu=F|_k\gamma_\nu. \tag{1.63}$$

By (1.31) and (1.32), the functions do not depend on the choice of the representatives γ_j. Since for any $\gamma\in\Gamma$, the set $\gamma_1\gamma,\dots,\gamma_\nu\gamma$ is again a set of representatives for the cosets, it follows that the functions $F_1|_k\gamma,\dots,F_\nu|_k\gamma$ coincide up to a permutation with the functions (1.63). By Lemma 1.22, each of the functions (1.63) belongs to $\mathfrak{M}$, and by Proposition 1.25, is a cusp form if F is a cusp form.

Let us now derive from Lemma 1.31 its generalization to the subspace $\mathfrak{N}=\mathfrak{N}_k(\Gamma(q))$ of cusp forms of $\mathfrak{M}$ in the following form: if

$$F(Z)=\sum_{A\in\mathbb{E},\,A>0}f(A)e^{\pi i\sigma(AZ)}\in\mathfrak{N},$$

and the Fourier coefficients $f(A)$ satisfy

$$f(A)=0\qquad\text{if}\quad\sigma(A)\le\frac{knq\nu}{2\pi b_n}, \tag{1.64}$$

then F is identically equal to zero. For that we shall consider the product

$$G(Z) = \prod_{j=1}^{\nu} F_j(Z)$$

of the functions (1.63). By (1.29) and the above considerations, we have, for every $M \in \Gamma$,

$$G|_{k\nu}M = j(M,Z)^{-k\nu} G(M\langle Z\rangle) = \prod_{j=1}^{\nu} (j(M,Z)^{-k} F_j(\langle Z\rangle))$$

$$= \prod_{j=1}^{\nu} F_j|_k M = \prod_{j=1}^{\nu} F_j = G.$$

Since G obviously satisfies the analytic conditions of the definition of cusp forms, we conclude that G is a cusp form of weight $k\nu$ for the group Γ. Let $f_j(A)$ be the Fourier coefficients of the function F_j, so that

$$F_j = \sum_{A \in \mathbb{E},\, A>0} f_j(A) e^{\frac{\pi i}{q}\sigma(AZ)}.$$

Then the Fourier coefficients $g(A)$ of G can be written in the form

$$g(A) = \sum_{A_1+\cdots+A_\nu = qA} f_1(A_1)\cdots f_\nu(A_\nu).$$

Let A be a positive definite matrix of $\mathbb{E}$ satisfying $\sigma(A) \le kn\nu/2\pi b$, where $b = b_n$. Then the inequality

$$\sigma(A) = \frac{1}{q}(\sigma(A_1) + \cdots + \sigma(A_\nu)) \le \frac{k\nu n}{2\pi b}$$

for positive definite even matrices $A_1, \ldots, A_\nu$ implies that $\sigma(A_j) < knq\nu/2\pi b$ for each $j = 1, \ldots, \nu$, since the trace of any positive definite matrix is positive. If, for example, $F_1 = F$ and so $f_1 = f$, then the condition (1.64) implies that each of the terms of the sum for $g(A)$ has a factor of the form $f_1(A_1) = f(A_1)$ with $\sigma(A_1) < knq\nu/2\pi b$, which is zero. Then by Lemma 1.29, $G = 0$, and so $F = 0$.

Now we can prove that the subspace $\mathfrak{N}$ of cusp forms of $\mathfrak{M}$ is finite-dimensional. Since entries of positive semidefinite matrices $A = (a_{\alpha\beta})$ satisfy the inequalities $a_{\alpha\alpha} \pm 2a_{\alpha\beta} + a_{\beta\beta} \ge 0$, it follows that the number of positive semidefinite even matrices A of order n with $\sigma(A) \le 2N$ does not exceed the bound

$$(N+1)^n(2N+1)^{n(n-1)/2}. \tag{1.65}$$

Therefore, the number of positive definite even matrices A satisfying the inequality in (1.64) is bounded by a number of the form $d_n(kq\nu)^{n(n+1)/2}$, where d_n depends only on n. Taking d_n to be integral, we see that $d+1$ arbitrary functions

$F_1,\dots,F_{d+1}$ of $\mathfrak{N}$ are linearly dependent, since one can always find complex numbers $c_1,\dots,c_{d+1}$ not all equal zero and such that the function $F = c_1F_1 + \cdots + c_{d+1}F_{d+1}$ satisfies the condition (1.64) and therefore is identically equal to zero.

Finally, we use induction on n to prove the theorem for the entire spaces $\mathfrak{M}^n = \mathfrak{M}_k(\Gamma^n(q))$. Let us define for $n \geq 1$ the linear map

$$\Phi = \Phi_n : \quad \mathfrak{M}^n \mapsto \underbrace{\mathfrak{M}^{n-1} \times \cdots \times \mathfrak{M}^{n-1}}_{\nu \text{ times}}$$

by

$$F|\Phi = (F_1|\Phi,\dots,F_\nu|\Phi) \qquad (F \in \mathfrak{M}^n),$$

where ν is the index of $\Gamma^n(q)$ in Γ^n, Φ the Siegel operator (1.56), F_j the functions (1.63), and where we set $\mathfrak{M}^0 = \mathbb{C}$. By Lemma 1.29, the kernel of Φ_n coincides with the subspace $\mathfrak{N}^n$ of cusp forms of $\mathfrak{M}^n$ and so is finite-dimensional. If the theorem is already proved for $n-1 \geq 1$, then the image of Φ_n is finite-dimensional and so is the space $\mathfrak{M}^n$. The image of Φ_1 is obviously finite-dimensional, which proves the theorem for $n = 1$. □

One can show that $\mathfrak{M}(S,\chi) = \{0\}$ if k is a negative integer, but we consider only modular forms of nonnegative integral weights and do not use this result.

Exercise 1.32. Show that $\mathfrak{M}_k(\Gamma^n) = \{0\}$ if nk is odd.

Exercise 1.33. Show that the spaces $\mathfrak{M}_k(\Gamma^1)$ for $k = 0,2,4,6,8,10$ contain no cusp form, and there is not more than one linearly independent cusp form of weight 12 for Γ^1.

[Hint: Use Lemma 1.31 for $n = 1$ with b_1 given in Exercise 1.17.]

Exercise 1.34. let k be a positive even integer. Show that the Fourier coefficients of any modular form

$$F(z) = \sum_{a=0}^{\infty} f(a)e^{2\pi iaz} \in \mathfrak{M}_k(\Gamma^1)$$

satisfy

$$f(a) = f(0)\zeta(k)^{-1}\frac{(2\pi i)^k}{(k-1)!}\sum_{d|n} d^{k-1} + f'(a), \quad \text{where } |f'(a)| \leq c_F a^{k/2},$$

where $\zeta(s)$ is the Riemann zeta function.

[Hint: Show first that the function $F(z) - f(0)(2\zeta(k))^{-1}E_k(z)$ is a cusp form. Then use Exercise 1.33 and (1.46).]

Exercise 1.35. Show that the spaces $\mathfrak{M}_k(\Gamma^1)$ for $k = 0,4,6,8,10$ are spanned respectively by 1, E_4, E_6, E_8, and E_{10}, where E_k are the Eisenstein series of Exercise 1.20.

Exercise 1.36. Show that the function

$$\Delta'(z) = ((2\zeta(4))^{-1}E_4(z))^3 - ((2\zeta(6))^{-1}E_6(z))^2$$

is a nonzero cusp form of the space $\mathfrak{M}_{12}(\Gamma^1)$ and that $\Delta'(z)$ and $E_{12}(z)$ span the space.

[Hint: With the help of classical formulas for $\zeta(4)$ and $\zeta(6)$ and formulas for Fourier coefficients of $E_4(z)$ and $E_6(z)$ show that the coefficient of $e^{2\pi iz}$ in the Fourier expansion of $\Delta'(z)$ is equal to 1728.]

Petersson Scalar Product. Every space $\mathfrak{N}_k(K,\chi)$ of cusp forms of an integral weight k and a congruence character χ for a congruence subgroup K of the symplectic group can be endowed with the structure of a Hilbert space by means of the scalar product. For two functions F and F' on $\mathbb{H} = \mathbb{H}^n$, we consider the differential form on $\mathbb{H}$ defined by

$$\omega_k(F,F') = F(Z)\overline{F'}(Z)h(Z)^k d^*Z, \tag{1.66}$$

where $h(Z) = h(X+iY) = \det Y$ is the height of Z, d^*Z is the invariant element of volume (1.12), and as usual, bar means complex conjugation. It follows from Lemma 1.4 and Proposition 1.6 that for each matrix $M = \left(\begin{smallmatrix} A & B \\ C & D \end{smallmatrix}\right) \in \mathbb{G} = \mathbb{G}^n$, the form satisfies the relation

$$\begin{aligned}
\omega_k(F,F')(M\langle Z\rangle) &= F(M\langle Z\rangle)\overline{F'}(M\langle Z\rangle)h(M\langle Z\rangle)^k d^*M\langle Z\rangle \\
&= \mu(M)^{nk}\det(CZ+D)^{-k}F(M\langle Z\rangle) \\
&\quad \times \det(C\overline{Z}+D)^{-k}\overline{F'}(M\langle Z\rangle)h(Z)^k d^*Z \\
&= \mu(M)^{-nk+n(n+1)}\left(F|_kM\right)(Z)\left(\overline{F'}|_kM\right)(Z)h(Z)^k d^*Z \\
&= \mu(M)^{n(n+1-k)}\omega_k\left(F|_kM, F'|_kM\right)(Z),
\end{aligned} \tag{1.67}$$

where $|_kM$ is the Petersson operator (1.29) of weight k. In particular, if $F, F' \in \mathfrak{M}_k(K,\chi)$ and $M \in K$, then by (1.32), we have

$$\omega_k(F,F')(M\langle Z\rangle) = \omega_k(\chi(M)F, \chi(M)F')(Z) = \omega_k(F,F')(Z). \tag{1.68}$$

It follows that the integral

$$\int_{D(K)} \omega_k(F,F')(Z) \tag{1.69}$$

on a fundamental domain $D(K)$ of K on $\mathbb{H}$ does not depend on the choice of the fundamental domain, provided that it converges absolutely.

Lemma 1.37. *Let K be a congruence subgroup of $\mathbb{S} = \mathbb{S}^n$, χ a congruence character of K, and k an integer. Suppose that at least one of the forms $F, F' \in \mathfrak{M}_k(K,\chi)$ is a cusp form. Then the integral* (1.69) *converges absolutely.*

Proof. By increasing the space $\mathfrak{M}_k(K,\chi)$, one may assume that K is a subgroup of finite index in Γ and the character χ is trivial on K. Let us take then as $D(K)$ a fundamental domain of the form (1.28) with $M = 1_{2n}$, that is, a finite union of the sets $\gamma_j\langle \mathbf{D}\rangle$ with $\gamma_j \in \Gamma$, where $\mathbf{D} = \mathbf{D}_n$ is the fundamental domain of Γ described in Theorem 1.16. Then, by (1.67), it is sufficient to show that each integral

$$\int_{\gamma\langle \mathbf{D}\rangle} \omega_k(F, F')(Z) = \int_{\mathbf{D}} \omega_k\left(F|_k\gamma, F'|_k\gamma\right)(Z) \text{ with } \gamma \in \Gamma$$

converges absolutely. Assuming that F' is a cusp form, by (1.42) and (1.43), we can write absolutely convergent Fourier expansions

$$F|_k\gamma = \sum_{A\in\mathbb{E}, A\geq 0} f_\gamma(A) e^{\frac{\pi i}{q}\sigma(AZ)}, \quad F'|_k\gamma = \sum_{A\in\mathbb{E}, A>.0} f'_\gamma(A) e^{\frac{\pi i}{q}\sigma(AZ)},$$

where $\mathbb{E} = \mathbb{E}^n$ and q is such that $\Gamma(q) \subset K$. It follows that

$$|(F|_k\gamma)(Z)(\overline{F'}|_k\gamma)(Z)| \leq \sum_{A\in\mathbb{E}, A>0} c(A) e^{-\frac{\pi}{q}\sigma(AZ)},$$

where $Z = X + iY$, with nonnegative coefficients $c(A)$, and the last series converges on $\mathbb{H}$. Then, by Theorem 1.16 and Lemma 1.24, we get the inequality

$$|(F|_k\gamma)(Z)(\overline{F'}|_k\gamma)(Z)| \leq d' e^{-d''\sigma(Y)} \quad (Z = X + iY \in \mathbf{D}),$$

with positive constants d' and d''. Thus, in order to prove the lemma, it suffices to show that the integral

$$\int_{\mathbf{D}} e^{-d\sigma(Y)} (\det Y)^{k-n-1} \prod_{1\leq\alpha\leq\beta\leq n} dx_{\alpha\beta} dy_{\alpha\beta}$$

with $d > 0$ converges. If $(x_{\alpha\beta}) + i(y_{\alpha\beta}) \in \mathbf{D}$, then it follows from the definition of $\mathbf{D}$ and (1.22) that $|x_{\alpha\beta}| \leq 1/2$ for $1 \leq \alpha, \beta \leq n$ and $y_{\alpha\beta} \leq y_{\alpha\alpha}/2$ for $\alpha \neq \beta$. In the proof of Theorem 1.16 we have seen that $y_{\alpha\alpha} \geq \sqrt{3}/2$ for $\alpha = 1, \ldots, n$. Then the inequality (1.22), if $k < n$, and the inequality (1.51), if $k \geq n+1$, imply that the last integral is majorized by

$$c \int_{\substack{|x_{\alpha\beta}|\leq 1/2, y_{\alpha\alpha}\geq\sqrt{3}/2, \\ |y_{\alpha\beta}|\leq y_{\alpha\alpha}/2 (\alpha\neq\beta)}} \prod_{\alpha=1}^{n} y_{\alpha\alpha}^{k-n-1} e^{-\delta y_{\alpha\alpha}} \prod_{1\leq\alpha\leq\beta\leq n} dx_{\alpha\beta} dy_{\alpha\beta}$$
$$= c \prod_{\alpha=1}^{\infty} \int_{\sqrt{3}/2}^{\infty} y_{\alpha\alpha}^{k-n-1+n-\alpha} e^{-\delta y_{\alpha\alpha}} dy_{\alpha\alpha} < \infty.$$

□

The above lemma justifies the following definition. For two modular forms $F, F' \in \mathfrak{M}(K, \chi)$ of an integral weight k and a congruence character χ for a congruence subgroup $K \subset \mathbb{S}$ such that at least one of the forms is a cusp form, the integral

$$(F,F')=\nu(\check{G})^{-1}\int_{D(G)}\omega_k(F,F')(Z), \tag{1.70}$$

where G is a congruence subgroup of Γ contained in K, $\check{G}=G\cup(-1_{2n})G$, $\nu(\check{G})$ the index of $\check{G}$ in Γ, and $D(G)$ a fundamental domain of G on $\mathbb{H}$, is called the *Petersson scalar product* of these forms.

Theorem 1.38. *Under the above assumption, the Petersson scalar product has the following properties:*

(1) *It converges absolutely and does not depend on the choice of fundamental domain $D(G)$.*
(2) *The scalar product is independent of the choice of the subgroup G of Γ contained in K.*
(3) *It is linear in F and conjugate linear in F'.*
(4) *It satisfies $\overline{(F,F')}=(F',F)$.*
(5) *If F is a cusp form, then $(F,F)\geq 0$, and $(F,F)=0$ only if $F=0$.*
(6) *If $M\in\mathbb{G}^n\cap\mathbb{Q}^{2n}_{2n}$ is a symplectic matrix with entries in the field $\mathbb{Q}$ of rational numbers and positive multiplier, then*

$$(F|_kM,\,F'|_kM)=\mu(M)^{n(k-n-1)}(F,F'), \tag{1.71}$$

where the functions $F|_kM$ and $F'|_kM$ are considered as elements of the space $\mathfrak{M}_k\left(M^{-1}KM,\chi_M\right)$ with the character χ_M defined by (1.34).

Proof. The first property follows from (1.68) and Lemma 1.37.

In order to prove the second property, let us assume that G' is another congruence subgroup contained in K. Then on replacing G' by $G'\cap G$, one can assume that $G'\subset G$. Let

$$\Gamma=\bigcup_i \check{G}\gamma_i \quad\text{and}\quad \check{G}=\bigcup_j \check{G}'\delta_j$$

be decompositions into different left cosets. Then we have

$$\Gamma=\bigcup_{i,j}\check{G}'\delta_j\gamma_i,$$

and the left cosets are distinct. It follows from Theorem 1.19 that one can take

$$D(G')=\bigcup_{i,j}(\delta_j\gamma_i)\langle\mathbf{D}_n\rangle,$$

hence we obtain

$$\begin{aligned}\nu(\check{G}')^{-1}\int_{D(G')}\omega_k(F,F')(Z)&=\nu(\check{G}')^{-1}\sum_i\sum_j\int_{\delta_j\langle\gamma_i\langle\mathbf{D}_n\rangle\rangle}\omega_k(F,F')(Z)\\&=\nu(\check{G}')^{-1}[\check{G}:\check{G}']\sum_i\int_{\gamma_i\langle\mathbf{D}_n\rangle}\omega_k(F,F')(Z),\end{aligned}$$

where we have also used relations (1.68). Again by Theorem 1.19, the last expression is equal to

$$\nu(\check{G})^{-1}\int_{D(G)}\omega_k(F,F')(Z),$$

which proves the property (2).

Properties (3), (4), and (5) follow directly from the definition.

In order to prove relation (1.71), we note that since all entries of M are rational, the group $M^{-1}KM$ is again a congruence subgroup of $\mathbb{S}$, and so the group $G_{(M)} = G\bigcap M^{-1}GM$ is again a congruence subgroup of Γ and, in particular, has finite index in Γ. It follows from Lemma 1.20, part (2), and (1.67) that

$$\begin{aligned}(F|_kM,F'|_kM) &= \nu\left(\check{G}_{(M)}\right)^{-1}\int_{D(G_{(M)})}\omega_k\left(F|_kM,F'|_kM\right)(Z)\\ &= \mu(M)^{n(k-n-1)}\nu\left(\check{G}_{(M)}\right)^{-1}\int_{D(G_{(M)})}\omega_k(F,F')\left(M\langle Z\rangle\right)\\ &= \mu(M)^{n(k-n-1)}\nu\left(\check{G}_{(M)}\right)^{-1}\int_{M\langle D(G_{(M)})\rangle}\omega_k(F,F')(Z),\end{aligned}$$

where $D\left(\check{G}_{(M)}\right)$ is a fundamental domain for $G_{(M)}$. It is clear that the set $M\langle D\left(G_{(M)}\right)\rangle$ is a fundamental domain for the group $MG_{(M)}M^{-1} = MGM^{-1}\cap G = G_{(M^{-1})}$. Hence, again by property (2), we can rewrite the last expression in the form

$$\begin{aligned}&\mu(M)^{n(k-n-1)}\nu\left(\check{G}_{(M)}\right)^{-1}\nu\left(\check{G}_{(M^{-1})}\right)\nu\left(\check{G}_{(M^{-1})}\right)^{-1}\int_{D\left(G_{(M^{-1})}\right)}\omega_k(F,F')(Z)\\ &= \mu(M)^{n(k-n-1)}\nu\left(\check{G}_{(M)}\right)^{-1}\nu\left(\check{G}_{(M^{-1})}\right)(F,F'),\end{aligned}$$

In order to prove the relation (1.71), it is sufficient to prove the equality

$$\nu\left(\check{G}_{(M)}\right) = \nu\left(\check{G}_{(M^{-1})}\right), \tag{1.72}$$

which follows from the following more general lemma.

Lemma 1.39. *Let K be a congruence subgroup of Γ^n, and let M be a matrix of $\mathbb{G}^n$ with rational entries. Then the groups $K_{(M)} = K\cap M^{-1}KM$ and $K_{(M^{-1})} = K\cap MKM^{-1}$ are congruence subgroups of Γ^n with equal indices in K:*

$$\left[K:K_{(M)}\right] = \left[K:K_{(M^{-1})}\right]. \tag{1.73}$$

Proof of the lemma. By Proposition 1.21 these groups are intersections of congruence subgroups of $\mathbb{S}^n$ and hence themselves are congruence subgroups. Let $\boldsymbol{D}$ be a fundamental domain for the group $K_{(M)}$. Since $K_{(M^{-1})} = MK_{(M)}M^{-1}$, it follows that one can take the set $M\langle\boldsymbol{D}\rangle$ as a fundamental domain for $K_{(M^{-1})}$. Then Theorem 1.19 implies the following relations for the invariant volumes of the domains $\boldsymbol{D}$ and $M\langle\boldsymbol{D}\rangle$:

$$\nu(\boldsymbol{D}) = \left[\check{K} : \check{K}_{(M)}\right] \nu(D(K)), \quad \nu(M\langle \boldsymbol{D} \rangle) = \left[\check{K} : \check{K}_{(M^{-1})}\right] \nu(D(K)),$$

where $\check{K} = K \cup (-1_{2n})K$. Since $\nu(\boldsymbol{D}) = \nu(M\langle \boldsymbol{D} \rangle)$, this relation implies both the equalities (1.72) and (1.73), We have proved the lemma and the theorem. □

Exercise 1.40. Show that the Eisenstein series $E_k(z)$ defined in Exercise 1.32 is orthogonal to the space $\mathfrak{N}_k(\Gamma^1)$ of cusp forms of weight k for Γ^1.

[Hint: Note that

$$E_k(z) = (1 + (-1)^k)\zeta(k) \sum_{M \in \Gamma^1_0 \backslash \Gamma^1} j(M, z)^{-k},$$

where $j(M, z)$ is defined by (1.30) and $\Gamma^1_0 = \left\{\left(\begin{smallmatrix} a & b \\ 0 & d \end{smallmatrix}\right) \in \Gamma^1\right\}$, which allows one to rewrite the scalar product of $E_k(z)$ on a cusp form as integral over a fundamental domain for the group Γ^1_0, say $\{z = x + iy \in \mathbb{H}^1 \mid -1/2 \le x \le 1/2\}$.]

Chapter 2
Dirichlet Series of Modular Forms

2.1 Radial Dirichlet Series

A natural way to approach zeta functions of modular forms is based on consideration of Dirichlet series constructed by means of Fourier coefficients of the forms. As was indicated in the introduction, a right zeta function must have certain analytic properties and an Euler product factorization. Therefore, the choice of appropriate Dirichlet series is motivated by a possibility of their analytic investigation plus a close relation with Euler products. These two features do not necessarily go together.

Radial Series of Cusp Forms. Let us consider modular forms of the spaces

$$\mathfrak{N}_k^n(q,\chi) = \mathfrak{N}_k\left(\Gamma_0^n(q), \boldsymbol{\chi}\right) \tag{2.1}$$

of cusp forms of nonnegative integral weights k for the congruence groups

$$\Gamma_0^n(q) = \left\{ \begin{pmatrix} A & B \\ C & D \end{pmatrix} \in \Gamma^n \;\middle|\; C \equiv 0 \pmod{q} \right\} \tag{2.2}$$

of integral levels q with characters $\boldsymbol{\chi}$ of the form

$$\boldsymbol{\chi}\left(\begin{pmatrix} A & B \\ C & D \end{pmatrix}\right) = \chi(\det D) \qquad \left(\begin{pmatrix} A & B \\ C & D \end{pmatrix} \in \Gamma_0^n(q)\right), \tag{2.3}$$

where χ is a Dirichlet character modulo q. According to Theorem 1.21 and Proposition 1.23, each such form F has the Fourier expansion

$$F(Z) = \sum_{A \in \mathbb{E}^n, A > 0} f(A) e^{\pi i \sigma(AZ)}, \tag{2.4}$$

which converges absolutely on $\mathbb{H}^n$ and uniformly on subsets (1.37) and has Fourier coefficients $f(A)$ satisfying the relations

$$f({}^tVAV) = \chi(\det V)(\det V)^k f(A) \qquad (\text{for all } A \in \mathbb{E}^n \text{ and } V \in \Lambda^n) \tag{2.5}$$

A. Andrianov, *Introduction to Siegel Modular Forms and Dirichlet Series*, Universitext, DOI 10.1007/978-0-387-78753-4_2, and the estimate

$$|f(A)| \le c(\det A)^{k/2}, \tag{2.6}$$

where $c = c_F$ depends only on F.

If $n = 1$, the Fourier expansion can be written the form

$$F(z) = \sum_{m=1}^{\infty} f(m) e^{2\pi i m z},$$

and the simplest Dirichlet series associated to F is

$$R(s) = R(s; F) = \sum_{m=1}^{\infty} \frac{f(m)}{m^s}, \tag{2.7}$$

It converges absolutely and uniformly in right half-planes $\Re s \ge k/2 + 1 + \varepsilon$ with $\varepsilon > 0$ and uniquely determines the form F. Using the *Euler formula*

$$\int_0^{\infty} y^{s-1} e^{-\alpha y} dy = \Gamma(s)\alpha^{-s} \qquad (\alpha > 0, \operatorname{Re} s > 0), \tag{2.8}$$

where $\Gamma(s)$ is the gamma function, we can see that the series (2.7) can be presented in the half-planes $\Re s \ge k/2 + 1 + \varepsilon$ by the absolutely and uniformly convergent *Mellin integral*

$$\begin{aligned} \Phi(s; F) &= \int_0^{\infty} F(iy) y^{s-1} dy \\ &= \sum_{m=0}^{\infty} f(m) \int_0^{\infty} y^{s-1} e^{-2\pi m y} dy = (2\pi)^{-s}\Gamma(s) R(s; F). \end{aligned} \tag{2.9}$$

In the next section we shall see that the integral representation allows one to show that the function (2.9) has an analytic continuation over the whole s-plane as a holomorphic function and satisfies a functional equation. In Chapter 5 we shall see that the series (2.7) has close ties with Euler products.

If $n > 1$, there is no unique natural way to introduce Dirichlet series corresponding to modular forms. The simplest generalizations of series (2.7) to genera $n > 1$ are given by so-called *radial Dirichlet series* (*corresponding to the matrix "ray"* $A, 2A, \ldots$)

$$R(s) = R(s; F, A) = \sum_{m=1}^{\infty} \frac{f(mA)}{m^s} \tag{2.10}$$

of a cusp form F with the Fourier expansion (2.4), where A is a fixed even positive definite matrix of order n. The estimate (2.6) implies that each radial series converges absolutely and uniformly in the right half-planes

$$\Re s > nk/2 + 1 + \varepsilon \tag{2.11}$$

with $\varepsilon > 0$. For $n > 1$, the radial Dirichlet series does not necessarily determine the initial modular form F, since it includes only Fourier coefficients depending on matrices of the given *ray* $A, 2A, \ldots$ of positive definite even matrices and gives no information on Fourier coefficients with arguments belonging to other rays.

Integral Representation of Radial Series. In order to derive an analytic expression of a radial series through original modular forms, we consider, for a fixed nonzero even matrix A of order n, the hyperplane in the space of real symmetric matrices of order n given by

$$X(A) = \left\{ X = {}^tX \in \mathbb{R}^n_n \;\middle|\; \sigma(AX) = 0 \right\} \tag{2.12}$$

and the lattice in the hyperplane defined by

$$X_0(A) = X(A) \bigcap \mathbb{Z}^n_n. \tag{2.13}$$

For every even matrix A' of order n, the mapping

$$X \mapsto e^{\pi i \sigma(A'X)} \qquad (X \in X(A))$$

defines clearly a character of the compact quotient group $X(A)/X_0(A)$, and the character is trivial if and only if A' is a multiple of A, $A' = mA$ (note that the corresponding hyperplanes and lattices coincide if $A' = mA$ and have nontrivial intersections otherwise). Thus,

$$\int_{X(A)/X_0(A)} e^{\pi i \sigma(A'X)} [dX] = \begin{cases} 1 & \text{if } A' = mA, \\ 0 & \text{otherwise,} \end{cases}$$

where $[dX]$ is the normalized Haar measure on $X(A)/X_0(A)$. Integrating termwise the Fourier expansion (2.4) of the restriction of F on subsets $X(A) + iY \subset \mathbb{H}^n$ over a fundamental domain $X(A)/X_0(A)$ of the translation group $X_0(A)$ on $X(A)$, we get

$$\begin{aligned} \int_{X(A)/X_0(A)} F(X+iY)[dX] &= \sum_{A' \in \mathbb{E}^n, A' > 0} f(A') e^{-\pi \sigma(A'Y)} \int_{X(A)/X_0(A)} e^{\pi i \sigma(A'X)} [dX] \\ &= \sum_{m \in \mathbb{Q}, mA \in \mathbb{E}^n, mA > 0} f(mA) e^{-\pi m \sigma(AY)}. \end{aligned}$$

We recall that the a *divisor* of nonzero even matrix A is the largest number $d \in \mathbb{N}$ such that the matrix $d^{-1}A$ is even. Nonzero even matrices of divisor $d = 1$ are called *primitive*. If A is even and primitive, then its rational multiple mA is even if and only if the coefficient m is integral. Since A and mA are both positive definite, m is positive. Thus, the last relation in the case of an even primitive positive definite matrix A turns into

$$\int_{X(A)/X_0(A)} F(X+iY)[dX] = \sum_{m=1}^{\infty} f(mA) e^{-\pi m \sigma(AY)}. \tag{2.14}$$

There are many ways to transform the power series on the right to corresponding Dirichlet series. For example, if we put in (2.14) $Y = v\sqrt{\det A}A^{-1}$ with $v > 0$, then, by using again the Euler integral (2.8), we obtain the identity

$$\int_0^\infty v^{s-1}\left(\int_{X(A)/X_0(A)} F(X + iv\sqrt{\det A}A^{-1})[dX]\right) dv$$
$$= \sum_{m=1}^\infty f(mA)\int_0^\infty v^{s-1}e^{-\pi nm\sqrt{\det A}v}dv = (\pi n\sqrt{\det A})^{-s}\Gamma(s)\sum_{m=1}^\infty \frac{f(mA)}{m^s}$$
$$= (\pi n\sqrt{\det A})^{-s}\Gamma(s)R(s;F,A) \tag{2.15}$$

In Section 2.2 we briefly recall the classical theory of Mellin transforms of cusp forms in one variable. The rest of this chapter is devoted to analytic properties of radial Dirichlet series of cusp forms for genus 2. In order to illustrate the main features of the problem, we consider in detail the case of cusp forms of level $q = 1$ and radial series corresponding to the matrix $A = \left(\begin{smallmatrix}2 & 0\\ 0 & 2\end{smallmatrix}\right)$ of the quadratic form $x^2 + y^2$. Some features of the situation for other A are presented in the exercises. At present, the radial series of cusp forms for genus 2 are sufficiently investigated only in the case of level $q = 1$ (see Notes). The problem of Euler factorization of the radial series for genera $n = 1$ and $n = 2$ will be treated in Chapter 5. If $n > 2$, both the problem of analytic investigation and the problem of Euler factorization of the radial series are still essentially open.

2.2 Mellin Transform of Cusp Forms in One Variable

Inversion Mapping. In this section we assume that $n = 1$. For a cusp form $F \in \mathfrak{N} = \mathfrak{N}_k^1(q,\chi)$ we consider the simplest radial Dirichlet series $R(s;F)$ defined by (2.7). As we have seen, in each of the right half-planes $\Re s \geq k/2 + 1 + \varepsilon$, the series can be presented by the absolutely and uniformly convergent Mellin integral (2.9).

Let us set

$$\Omega = \Omega^1(q) = \begin{pmatrix} 0 & -1 \\ q & 0 \end{pmatrix}. \tag{2.16}$$

Since

$$\Omega\begin{pmatrix} a & b \\ c & d \end{pmatrix}\Omega^{-1} = \begin{pmatrix} d & -c/q \\ -qb & a \end{pmatrix},$$

it follows that $\Omega K_0 \Omega^{-1} = K_0$, where $K_0 = \Gamma_0^1(q)$, and the function

$$F^*(z) = F|_k\Omega = q^{-1}z^{-k}F(-1/qz) \tag{2.17}$$

satisfies

$$F^*|_kM = F|_k\Omega M\Omega^{-1}|_k\Omega = \chi(a)F|_k\Omega = \bar{\chi}(d)F^*, \text{ for each } M = \begin{pmatrix} a & b \\ c & d \end{pmatrix} \in K_0,$$

where $\bar{\chi} = \chi^{-1}$ is the conjugate character. It is easy to see with the help of the Euclidean algorithm that among left multiples of any integral matrix of order 2 by matrices of $\Gamma = \Gamma^1 = \mathrm{SL}_2(\mathbb{Z})$ there are always upper triangular matrices. In particular, for every matrix $M \in \Gamma$ there exists a matrix $M' \in \Gamma$ such that

$$\Omega M = M' \begin{pmatrix} * & * \\ 0 & * \end{pmatrix}.$$

Since F is a cusp form, it follows that for every $M \in \Gamma$ the function $F^*|_k M$ has a Fourier expansion of the form (1.44). We conclude that $F^* \in \mathfrak{N}(q, \bar{\chi})$. Since

$$F|_k \Omega^2 = F|_k \begin{pmatrix} -q & 0 \\ 0 & -q \end{pmatrix} = (-1)^k q^{k-2} F,$$

the *inversion mapping* $F \mapsto F^*$ satisfies

$$(F^*)^* = (-1)^k q^{k-2} F \tag{2.18}$$

and so determines an isomorphism of the space $\mathfrak{N}(q, \chi) = \mathfrak{N}_k^1(q, \chi)$ onto $\mathfrak{N}(q, \bar{\chi}) = \mathfrak{N}_k^1(q, \bar{\chi})$. In particular, if $\chi = \bar{\chi}$, i.e., the character χ is real and takes only values ± 1, the mapping is an automorphism of the space $\mathfrak{N}(q, \chi)$. In this case the inversion mapping

$$F \mapsto (-i)^k q^{-k/2+1} F^* \qquad \text{with } i = \sqrt{-1}$$

is also an automorphism of the space, and its square is the identity map. It follows that in the case of real χ, one can write a direct sum decomposition

$$\mathfrak{N}(q, \chi) = \mathfrak{N}^+(q, \chi) + \mathfrak{N}^-(q, \chi), \tag{2.19}$$

where

$$(-i)^k q^{-k/2} z^{-k} F(-1/qz) = \pm F \qquad \text{if } F \in \mathfrak{N}^{\pm}(q, \chi). \tag{2.20}$$

Analytic Continuation and Functional Equation. Let us return to the integral (2.9).

Theorem 2.1. *For a nonnegative integer k, a positive integer q, and a Dirichlet character χ modulo q, let $F \in \mathfrak{N}(q, \chi) = \mathfrak{N}_k^1(q, \chi)$ be a cusp form of weight k and character χ of the form* (2.3) *for the group $K_0 = \Gamma_0^1(q)$. Then the function*

$$\Phi(s; F) = (2\pi)^{-s} \Gamma(s) R(s; F),$$

where $R(s; F)$ is the Dirichlet series (2.7), *has the following properties:*

(1) *The function $\Phi(s; F)$ has an analytic continuation over the whole s-plane as a holomorphic function.*
(2) *The function $\Phi(s; F)$ satisfies the functional equation*

$$\Phi(k - s; F) = i^k q^{s-k+1} \Phi(s; F^*), \tag{2.21}$$

where $F^* \in \mathfrak{N}(q, \bar{\chi})$ is the function (2.17).

(3) *If the character* χ *is real, and the form* F *belongs to one of the subspaces* $\mathfrak{N}^{\pm}(q,\chi)$ *of the decomposition* (2.19), *then the function* $\Phi(s;F)$ *satisfies*

$$\Phi(k-s;F)=\pm(-1)^k q^{s-k/2}\Phi(s;F). \tag{2.22}$$

Proof. Similarly to the case $\chi=1$ considered in the introduction, we can write, for $\Re s$ sufficiently large, the identity

$$\begin{aligned}\Phi(s;F)&=\int_0^{q^{-1/2}}F(iy)y^{s-1}dy+\int_{q^{-1/2}}^{\infty}F(iy)y^{s-1}dy\\&=\int_{q^{-1/2}}^{\infty}F(i/qy)(1/qy)^{s-1}(1/qy^2)dy+\int_{q^{-1/2}}^{\infty}F(iy)y^{s-1}dy,\end{aligned}$$

where in the first integral we have changed the variable $y\mapsto 1/qy$. By substituting $F(i/qy)=q(iy)^kF^*(iy)$ in the first integral, we can rewrite the last relation in the form

$$\Phi(s;F)=i^kq^{1-s}\int_{q^{-1/2}}^{\infty}F^*(iy)y^{k-s-1}dy+\int_{q^{-1/2}}^{\infty}F(iy)y^{s-1}dy. \tag{2.23}$$

Since F and F^* are both cusp forms, the functions $|F(iy)|$ and $|F^*(iy)|$ approach exponentially zero as $y\to\infty$ (see Lemma 1.24). It follows that the integrals in (2.23) both converge absolutely and uniformly on any compact subset of the s-plane. This proves part (1).

By (2.23) with $k-s$ in place of s and (2.18), we obtain

$$\begin{aligned}\Phi(k-s;F)&=i^kq^{s-k+1}\int_{q^{-1/2}}^{\infty}F^*(iy)y^{s-1}dy+\int_{q^{-1/2}}^{\infty}F(iy)y^{k-s-1}dy\\&=i^kq^{s-k+1}\Big(\int_{q^{-1/2}}^{\infty}F^*(iy)y^{s-1}dy\\&\qquad+(-i)^kq^{k-s-1}\int_{q^{-1/2}}^{\infty}(-1)^kq^{2-k}(F^*)^*(iy)y^{k-s-1}dy\Big)\\&=i^kq^{s-k+1}\Phi(s;F^*),\end{aligned}$$

which proves the functional equation (2.21).

Finally, if, say, $F\in\mathfrak{N}^{-}(q,\chi)$, then it satisfies $(-i)^kq^{-k/2+1}F^*=-F$, and so the functional equation turns into the equation

$$\Phi(k-s;F)=i^kq^{s-k+1}(-1)i^kq^{k/2-1}\Phi(s;F)=-(-1)^kq^{s-k/2}\Phi(s;F).$$

□

Exercise 2.2. Let $F\in\mathfrak{M}_k(q,\chi)$ be an arbitrary modular form with Fourier expansion

$$F(z)=f(0)+\sum_{m=1}^{\infty}f(m)e^{2\pi imz}.$$

Show that the Dirichlet series

$$R(s;F) = \sum_{m=1}^{\infty} \frac{f(m)}{m^s}$$

converge absolutely and uniformly in a right half-plane of the variable s; show that the functions $\Phi(s;F) = (2\pi)^{-s}\Gamma(s)R(s;F)$ can be continued analytically to the whole s-plane as a meromorphic function having at most two simple poles at points $s = 0$ and $s = k$ and satisfying the functional equation $\Phi(k-s;F) = i^k q^{s-k+1}\Phi(s;F^*)$, where $F^* \in \mathfrak{M}(q,\bar{\chi})$ is the function (2.17); find the residues at the poles of $\Phi(s;F)$.

[Hint: Use the integral representation $\Phi(s;F) = \int_0^\infty (F(it) - f(0))y^{s-1}dy$.]

Exercise 2.3. Compute explicitly in the terms of the Riemann zeta function the Dirichlet series $R(s;E_k)$ for the Eisenstein series $E_k(z)$ of even weight $k > 2$ defined in Exercise 1.32.

[Hint: Use Exercise 1.33.]

2.3 Transformations of Lobachevsky Half-Spaces

Transformation Group. The integral representation (2.15) for $n = 2$ shows that the radial Dirichlet series $R(s;F,A)$ can be obtained by integration of the restriction of F to a real three-dimensional domain

$$H(A) = X(A) + iv\sqrt{\det A}A^{-1} \subset \mathbb{H}^2. \tag{2.24}$$

Further transformation of this representation is based on the remarkable circumstance that the real symplectic group $\mathbb{S}^2 = \mathrm{Sp}_2(\mathbb{R})$ defined by (1.13) has a rather large subgroup $S(A)$ acting as a transitive group of analytic automorphisms of the domain $H(A)$. By the end of this chapter the genus n is equal to 2, and the corresponding index is usually omitted.

Let us consider now in detail the particular case $A = 2E = 21_2$, where we write E in place of 1_2. In this case one can take

$$\begin{cases} \mathbf{X} = X(2E) = \left\{ \begin{pmatrix} x & y \\ y & -x \end{pmatrix} \,\middle|\, x, y \in \mathbb{R} \right\}, \\ \mathbf{X}_0 = X_0(2E) = \left\{ \begin{pmatrix} x & y \\ y & -x \end{pmatrix} \,\middle|\, x, y \in \mathbb{Z} \right\}, \\ \mathbf{X}/\mathbf{X}_0 = X(2E)/X_0(2E) = \left\{ \begin{pmatrix} x & y \\ y & -x \end{pmatrix} \,\middle|\, |x| \le 1/2, |y| \le 1/2 \right\}. \end{cases} \tag{2.25}$$

Since clearly, one can take $[dX] = dx\,dy$, the identity (2.15) for $A = 2E$ turns into the identity

$$(4\pi)^{-s}\Gamma(s)R(s;F,2E)$$
$$= \int_0^\infty \left\{ \int_{\substack{|x|\le 1/2,\\ |y|\le 1/2}} F\left(\begin{pmatrix} x+iv & y \\ y & -x+iv \end{pmatrix}\right) dx\,dy \right\} v^{s-1}\,dv. \qquad (2.26)$$

The domain of integration of this double integral is contained in the open subset

$$\mathbf{H} = H(2E) = \left\{ \begin{pmatrix} x+iv & y \\ y & -x+iv \end{pmatrix} \,\middle|\, x,y,v \in \mathbb{R},\ v>0 \right\} \subset \mathbb{H}^2. \qquad (2.27)$$

Let us consider the set

$$\mathbf{S} = S(2E) = \left\{ \begin{pmatrix} A & B \\ C & D \end{pmatrix} \,\middle|\, A,D \in \mathbf{Y},\quad B,C \in \mathbf{X},\quad A^tD - B^tC = E \right\}, \qquad (2.28)$$

where

$$\mathbf{Y} = Y(2E) = \left\{ \begin{pmatrix} x & y \\ -y & x \end{pmatrix} \,\middle|\, x,y \in \mathbb{R} \right\}. \qquad (2.29)$$

Proposition 2.4. *The set* $\mathbf{S}$ *is a subgroup of the group* $\mathbb{S} = Sp_2(\mathbb{R})$. *For every matrix* $M = \left(\begin{smallmatrix} A & B \\ C & D \end{smallmatrix}\right) \in \mathbf{S}$, *the transformation*

$$Z \mapsto M\langle Z\rangle = (AZ+B)(CZ+D)^{-1} \qquad (Z \in \mathbf{H})$$

is a real analytic automorphism of the domain $\mathbf{H}$. *The action of* $\mathbf{S}$ *on* $\mathbf{H}$ *is transitive.*

Proof. Since a matrix of the form $\left(\begin{smallmatrix} x & y \\ -y & x \end{smallmatrix}\right)\begin{pmatrix} x' & y' \\ y' & -x' \end{pmatrix}$ is always symmetric, it follows from (1.3) and (1.13) that $\mathbf{S} \subset \mathbb{S}$. Let us set

$$\mathbf{R} = \left\{ \begin{pmatrix} A & B \\ C & D \end{pmatrix} \,\middle|\, A,D \in \mathbf{Y},\quad B,C \in \mathbf{X} \right\}.$$

Then $\mathbf{S} = \mathbb{S} \cap \mathbf{R}$. If $M, M' \in \mathbf{S}$, then $M, M' \in \mathbb{S}$ and so $MM' \in \mathbb{S}$. From the obvious fact that $\mathbf{X}$ and $\mathbf{Y}$ are abelian groups under the addition of matrices, and from easily verified inclusions

$$\mathbf{YY} \subset \mathbf{Y}, \quad \mathbf{XX} \subset \mathbf{X}, \quad \mathbf{XY} \subset \mathbf{X}, \quad \mathbf{YX} \subset \mathbf{X}$$

we conclude that $MM' \in \mathbf{R}$. Thus, $MM' \in \mathbb{S} \cap \mathbf{R} = \mathbf{S}$. Further, the formula (1.1) implies that $M^{-1} \in \mathbf{S}$ if $M \in \mathbf{S}$. Hence $\mathbf{S}$ is a subgroup of $\mathbb{S}$.

Now we shall show that

$$\mathbf{H} = \left\{ M\langle iE\rangle \,\middle|\, M \in \mathbf{S} \right\}. \qquad (2.30)$$

Let $Z = \begin{pmatrix} x & y \\ y & -x \end{pmatrix} + iv \in \mathbf{H}$. It is clear that the matrix

$$M_Z = \begin{pmatrix} \sqrt{v}E & \frac{1}{\sqrt{v}}\begin{pmatrix} x & y \\ y & -x \end{pmatrix} \\ 0 & \frac{1}{\sqrt{v}} \end{pmatrix} \quad \text{is in } \mathbf{S} \tag{2.31}$$

and $M_Z\langle iE\rangle = Z$. Thus, the left-hand side of (2.30) is contained in the right-hand side. In order to prove the reverse inclusion we note that if $M = \begin{pmatrix} A & B \\ C & D \end{pmatrix} \in \mathbf{S}$ with

$$A = \begin{pmatrix} x_1 & y_1 \\ -y_1 & x_1 \end{pmatrix}, \quad B = \begin{pmatrix} x_2 & y_2 \\ y_2 & -x_2 \end{pmatrix}, \quad C = \begin{pmatrix} x_3 & y_3 \\ y_3 & -x_3 \end{pmatrix}, \quad D = \begin{pmatrix} x_4 & y_4 \\ -y_4 & x_4 \end{pmatrix},$$

and $M\langle iE\rangle = X' + iY'$, then, by (1.10) and (1.3), we have

$$\begin{aligned} Y' &= {}^t(D - iC)^{-1}(D + iC)^{-1} = (D^tD + C^tC)^{-1} \\ &= \left(x_3^2 + x_4^2 + y_3^2 + y_4^2\right)^{-1} E. \end{aligned} \tag{2.32}$$

Thus, it is sufficient to check that $\sigma(X') = 0$. Since $\det(D + iC) = x_3^2 + y_3^2 + x_4^2 + y_4^2$, we have

$$\begin{aligned} M\langle iE\rangle &= \left(x_3^2 + y_3^2 + x_4^2 + y_4^2\right)^{-1} \\ &\times \begin{pmatrix} x_2 + ix_1 & y_2 + iy_1 \\ y_2 - iy_1 & -x_2 + ix_1 \end{pmatrix}\begin{pmatrix} x_4 - ix_3 & -y_4 - iy_3 \\ y_4 - iy_3 & x_4 + ix_3 \end{pmatrix}, \end{aligned} \tag{2.33}$$

hence

$$\left(x_3^2 + y_3^2 + x_4^2 + y_4^2\right)\sigma(X') = x_2x_4 + x_1x_3 + y_2y_4 + y_1y_3 - y_2y_4 - y_1y_3 - x_2x_4 - x_1x_3 = 0.$$

We have proved the relation (2.30), which implies the assertions of the proposition on the action of $\mathbf{S}$ on $\mathbf{H}$. □

The above proposition shows that the domain $\mathbf{H} \subset \mathbb{H}^2$ can be considered as a homogeneous space for the Lie group $\mathbf{S} \subset \mathbb{S}^2$. Next, we shall show that the pair $(\mathbf{S}, \mathbf{H})$ is naturally isomorphic to the pair $(\mathbf{G}, \mathbf{L})$, where $\mathbf{G} = \mathrm{SL}_2(\mathbb{C})$, $\mathbf{L}$ is the 3-dimensional hyperbolic half-space (the *Lobachevsky space*)

$$\mathbf{L} = \left\{u = (z, v) \,\middle|\, z = x + iy \in \mathbb{C}, \quad v > 0\right\},$$

and the action of $\mathbf{G}$ on $\mathbf{L}$ is given by the rule

$$\mathbf{G} \ni g = \begin{pmatrix} \alpha & \beta \\ \gamma & \delta \end{pmatrix}:$$

$$u = (z, v) \mapsto g(u) = \left(\frac{(\alpha z + \beta)(\bar{\gamma}\bar{z} + \bar{\delta}) + \alpha\bar{\gamma}v^2}{\Delta(g, u)}, \frac{v}{\Delta(g, u)}\right), \tag{2.34}$$

where the bar means complex conjugation and where

$$\Delta(g, u) = |\gamma z + \delta|^2 + |\gamma|^2 v^2. \tag{2.35}$$

In order to establish the relation of the pairs $(\mathbf{S}, \mathbf{H})$ and $(\mathbf{G}, \mathbf{L})$ for a real matrix of the form

$$M = \begin{pmatrix} A & B \\ C & D \end{pmatrix} = \begin{pmatrix} x_1 & y_1 & x_2 & y_2 \\ -y_1 & x_1 & y_2 & -x_2 \\ x_3 & y_3 & x_4 & y_4 \\ y_3 & -x_3 & -y_4 & x_4 \end{pmatrix} \in \mathbf{S}, \tag{2.36}$$

we set

$$g(M) = \begin{pmatrix} \alpha & \beta \\ \gamma & \delta \end{pmatrix} = \begin{pmatrix} x_1 + iy_1 & y_2 + ix_2 \\ y_3 - ix_3 & x_4 - iy_4 \end{pmatrix}, \tag{2.37}$$

and for

$$Z = \begin{pmatrix} x & y \\ y & -x \end{pmatrix} + ivE \in \mathbf{H} \tag{2.38}$$

we write

$$u(Z) = (y + ix,\, v) \in \mathbf{L}. \tag{2.39}$$

Proposition 2.5. *In the above notation the following assertions hold:*

(1) *$g(\mathbf{S}) = \mathbf{G}$, and the map $g : \mathbf{S} \mapsto \mathbf{G}$ is an isomorphism of real Lie groups; the map $u : \mathbf{H} \mapsto \mathbf{L}$ is a real analytic isomorphism.*

(2) *The maps g and u transform the action of $\mathbf{S}$ on $\mathbf{H}$ into the action* (2.34) *of $\mathbf{G}$ on $\mathbf{L}$ in the sense that*

$$u(M\langle Z\rangle) = g(M)\,(u(Z)) \qquad (M \in \mathbf{S}, Z \in \mathbf{H}); \tag{2.40}$$

in particular, the action (2.34) *satisfies*

$$(gg')(u) = g(g'(u)) \qquad (g, g' \subset \mathbf{G}, u \in \mathbf{L}). \tag{2.41}$$

(3) *For all $M = \begin{pmatrix} A & B \\ C & D \end{pmatrix} \in \mathbf{S}$ and $Z \in \mathbf{H}$, the identity*

$$\det(CZ + D) = \Delta(g(M), u(Z)) \tag{2.42}$$

holds; in particular, the function $(g, u) \mapsto \Delta(g, u)$ is a factor of automorphy of the pair $(\mathbf{G}, \mathbf{L})$, i.e., it does not vanish on $\mathbf{G} \times \mathbf{L}$ and satisfies

$$\Delta(gg', u) = \Delta(g, g'(u))\Delta(g', u) \quad (g, g' \in \mathbf{G}, u \in \mathbf{L}). \tag{2.43}$$

Proof. The condition $M \in \mathbf{S}$ for a real matrix of the form (2.36) is equivalent to the relation $A^tD - B^tC = E$, that is, the relations

$$x_1x_4 + y_1y_4 - x_2x_3 - y_2y_3 = 1, \quad -y_1x_4 + x_1y_4 - y_2x_3 + x_2y_3 = 0,$$

which can be rewritten in the form $(x_1+iy_1)(x_4-iy_4)-(y_2+ix_2)(y_3-ix_3)=1$. The last relation is equivalent to the inclusion $g(M)\in \mathrm{SL}_2(\mathbb{C})=\mathbf{G}$. The relations $g(MM')=g(M)g(M')$ for matrices M and M' of the form (2.36) follow from the definitions by direct multiplication of the matrices, which we leave to the reader. The rest of part (1) is clear.

In order to prove the relations (2.40), let us set

$$MM_Z = M' = \begin{pmatrix} A' & B' \\ C' & D' \end{pmatrix},$$

where M_Z is the matrix (2.31). By multiplying the matrices, we get

$$A' = \begin{pmatrix} x_1' & y_1' \\ -y_1' & x_1' \end{pmatrix} = \sqrt{v}\begin{pmatrix} x_1 & y_1 \\ -y_1 & x_1 \end{pmatrix},$$

$$B' = \begin{pmatrix} x_2' & y_2' \\ y_2' & -x_2' \end{pmatrix} = \frac{1}{\sqrt{v}}\begin{pmatrix} x_1x+y_1y+x_2 & x_1y-y_1x+y_2 \\ -y_1x+x_1y+y_2 & -y_1y-x_1x-x_2 \end{pmatrix},$$

$$C' = \begin{pmatrix} x_3' & y_3' \\ y_3' & -x_3' \end{pmatrix} = \sqrt{v}\begin{pmatrix} x_3 & y_3 \\ y_3 & -x_3 \end{pmatrix},$$

$$D' = \begin{pmatrix} x_4' & y_4' \\ -y_4' & x_4' \end{pmatrix} = \frac{1}{\sqrt{v}}\begin{pmatrix} x_3x+y_3y+x_4 & x_3y-y_3x+y_4 \\ y_3x-x_3y-y_4 & y_3y+x_3x+x_4 \end{pmatrix}.$$

For the matrix

$$X'+iY' = M\langle Z\rangle = M\langle M_Z\langle iE\rangle\rangle = M'\langle iE\rangle,$$

by the identities (2.32) and (2.33) with M' in place of M, we obtain, respectively,

$$Y' = \left((x_3')^2+(y_3')^2+(x_4')^2+(y_4')^2\right)^{-1}E$$

and

$$X'+iY' = \left((x_3')^2+(y_3')^2+(x_4')^2+(y_4')^2\right)^{-1} \times \begin{pmatrix} x_2'+ix_1' & y_2'+iy_1' \\ y_2'-iy_1' & -x_2'+ix_1' \end{pmatrix}\begin{pmatrix} x_4'-ix_3' & -y_4'-iy_3' \\ y_4'-iy_3' & x_4'+ix_3' \end{pmatrix},$$

hence

$$X' = \left((x_3')^2+(y_3')^2+(x_4')^2+(y_4')^2\right)^{-1}\begin{pmatrix} x' & y' \\ y' & -x' \end{pmatrix}$$

with $x' = x_2'x_4'+x_1'x_3'+y_2'y_4'+y_1'y_3'$ and $y' = -x_2'y_4'+x_1'y_3'+y_2'x_4'-y_1'x_3'$. By substituting, we get

$$(x_3')^2+(y_3')^2+(x_4')^2+(y_4')^2 = v(x_3^2+y_3^2)+\frac{1}{v}\left((x_3x+y_3y+x_4)^2+(x_3y-y_3x+y_4)^2\right)$$
$$= \frac{1}{v}\left(v^2|\gamma|^2+|\gamma(y+ix)+\delta|^2\right) = \frac{\Delta(g,u)}{v},$$

where $g = g(M)$ and $u = u(Z)$ are given by (2.37) and (2.39), and

$$\begin{aligned}
y' + ix' &= (-x_2'y_4' + x_1'y_3' + y_2'x_4' - y_1'x_3') + i(x_2'x_4' + x_1'x_3' + y_2'y_4' + y_1'y_3') \\
&= \Big(-\frac{1}{v}(x_1x + y_1y + x_2)(x_3y - y_3x + y_4) + vx_1y_3 \\
&\quad + \frac{1}{v}(x_1y - y_1x + y_2)(x_3x + y_3y + x_4) - vy_1x_3 \Big) \\
&\quad + i\Big(\frac{1}{v}(x_1x + y_1y + x_2)(x_3x + y_3y + x_4) + vx_1x_3 \\
&\quad + \frac{1}{v}(x_1y - y_1x + y_2)(x_3y - y_3x + y_4) + vy_1y_3 \Big) \\
&= \frac{1}{v}((x_1 + iy_1)(y + ix) + (y_2 + ix_2))((y_3 + ix_3)(y - ix) + (x_4 + iy_4)) \\
&\quad + v(x_1 + iy_1)(y_3 + ix_3) = \frac{1}{v}(\alpha z + \beta)(\bar{\gamma}\bar{z} + \bar{\delta}) + \alpha\bar{\gamma}v.
\end{aligned}$$

The above computations imply the formula (2.40):

$$\begin{aligned}
u(M\langle Z\rangle) &= u(X' + iY') = u\left(\frac{v}{\Delta(g,u)} \begin{pmatrix} x' & y' \\ y' & -x' \end{pmatrix} + i\frac{v}{\Delta(g,u)}E \right) \\
&= \left(\frac{v}{\Delta(g,u)}(y' + ix'), \frac{v}{\Delta(g,u)} \right) = g(M)\,(u(Z)).
\end{aligned}$$

Formula (2.41) follows from (2.40) and (1.7).

Direct computation shows that

$$\begin{aligned}
\det(CZ + D) &= \det\left(\begin{pmatrix} x_3 & y_3 \\ y_3 & -x_3 \end{pmatrix} \begin{pmatrix} x + iv & y \\ y & -x + iv \end{pmatrix} + \begin{pmatrix} x_4 & y_4 \\ -y_4 & x_4 \end{pmatrix} \right) \\
&= (x_3x + y_3y + x_4)^2 - (ivx_3)^2 + (x_3y - y_3x + y_4)^2 - (ivy_3)^2 \\
&= \Delta(g,u),
\end{aligned}$$

as we have seen above. The relation (2.43) follows from (2.42) and (1.9). □

For further integration on **L** we shall need a **G**-invariant volume element.

Proposition 2.6. *The volume element on* **L** *given by*

$$du = \frac{dx\,dy\,dv}{v^3} \qquad (u = (x + iy, v) \in L) \tag{2.44}$$

is invariant under every transformation (2.34).

Proof. It is easy to see that the group $\mathbf{G} = \mathrm{SL}_2(\mathbb{C})$ is generated by the matrices

$$\begin{pmatrix} \alpha & 0 \\ 0 & \alpha^{-1} \end{pmatrix}, \quad \begin{pmatrix} 1 & \beta \\ 0 & 1 \end{pmatrix}, \quad \text{and} \quad \begin{pmatrix} 0 & 1 \\ -1 & 0 \end{pmatrix} \tag{2.45}$$

with $\alpha, \beta \in \mathbb{C}$, $\alpha \neq 0$. For each of the matrices (2.45), the check of the assertion is an easy exercise, which we leave to the reader. □

Exercise 2.7. Let A be a real symmetric positive definite matrix of order 2, and suppose $\tau \in \mathrm{GL}_2(\mathbb{R})$ satisfies $\tau A {}^t\tau = \sqrt{\det A}E$. Show that for every matrix M contained in the group

$$S(A) = M_\tau \mathbf{S} M_\tau^{-1} \qquad \text{with} \quad M_\tau = \begin{pmatrix} {}^t\tau & O \\ O & \tau^{-1} \end{pmatrix},$$

the transformation $Z \mapsto M\langle Z\rangle$ defines a real analytic automorphism of the domain H(A) given by (2.24) and (2.12), and the action of $S(A)$ on $H(A)$ is transitive.

[Hint: Show first that $H(A) = M_\tau\langle \mathbf{H}\rangle$.]

Exercise 2.8. In the notation and assumptions of the previous exercise show that Proposition 2.5 remains true if one replaces the pair **S**, **H** by the pair $S(A)$, $H(A)$, and the maps g, u by the maps $g_\tau : S(A) \mapsto \mathbf{G}$, $u_\tau : H(A) \mapsto \mathbf{L}$ defined by

$$g_\tau(M) = g\left(M_\tau^{-1} M M_\tau\right) \quad (M \in S(A)) \text{ and } u_\tau(Z) = u\left(M_\tau^{-1}\langle Z\rangle\right) \quad (Z \in H(A)).$$

Discrete Subgroup. We turn now to integration of restrictions of modular forms to Lobachevsky subspaces of $\mathbb{H}^2$. By the end of this chapter we shall restrict our consideration to the case of cusp forms for the group $\Gamma^2 = \Gamma_0^2(1)$ with trivial character.

Each function $F \in \mathfrak{N}_k^2 = \mathfrak{N}_k^2(1,1)$ satisfies the functional equations

$$\det(CZ+D)^{-k} F\left(M\langle Z\rangle\right) = F \qquad \text{for all } M = \begin{pmatrix} A & B \\ C & D \end{pmatrix} \in \Gamma^2. \tag{2.46}$$

Let $\widetilde{F}$ be the function on the Lobachevsky half-space **L** corresponding via (2.39) to the restriction of F on $\mathbf{H} \subset \mathbb{H}^2$, i.e.,

$$\widetilde{F}(u) = F(Z) \quad \text{if } u = u(Z) \text{ with } Z \in \mathbf{H}, \tag{2.47}$$

and let

$$g = \begin{pmatrix} \alpha & \beta \\ \gamma & \delta \end{pmatrix} = g(M) \quad \text{with} \quad M = \begin{pmatrix} A & B \\ C & D \end{pmatrix} \in \mathbf{S} \bigcap \Gamma^2. \tag{2.48}$$

Then the relations (2.46) and Proposition 2.6 imply the relations

$$\begin{aligned} \Delta(g,u)^{-k} \widetilde{F}(g(u)) &= \Delta(g(M), u(Z))^{-k} \widetilde{F}\left(g(M)(u(Z))\right) \\ &= \det(CZ+D)^{-k} F\left(M\langle Z\rangle\right) = F(Z) = \widetilde{F}(u). \end{aligned} \tag{2.49}$$

Lemma 2.9. *The restriction of the map $M \mapsto g(M)$ to the subgroup $\mathbf{S} \cap \Gamma^2$ is an isomorphism of the subgroup onto the special linear group*

$$\mathbf{\Lambda} = \Lambda(\mathcal{O}) = \mathrm{SL}_2(\mathcal{O}) \tag{2.50}$$

of order 2 over the ring of Gaussian integers $\mathcal{O} = \mathbb{Z} + i\mathbb{Z}$, i.e., the ring of integers of the imaginary quadratic field $\mathbb{Q}[i]$ $(i = \sqrt{-1})$.

Proof. Since the map g is isomorphism of $\mathbf{S}$ on $SL_2(\mathbb{C})$, it is sufficient to note that a matrix M of the form (2.36) contained in $\mathbf{S}$ belongs to the group Γ^2 if and only if all of its entries belong to $\mathbb{Z}$. This is equivalent to the conditions that $g(M) = \begin{pmatrix} \alpha & \beta \\ \gamma & \delta \end{pmatrix} \in SL_2(\mathbb{C})$ with α, β, γ, δ belonging to the ring $\mathbb{Z}+i\mathbb{Z} = \mathcal{O}$. □

Analogously to the case of the symplectic modular group considered in Section 1.2, the group $\mathbf{\Lambda}$ is a discrete subgroups of $\mathbf{G}$ discretely acting on $\mathbf{L}$. Similarly to Theorem 1.16, we have the following theorem.

Theorem 2.10. *The closed subset of* $\mathbf{L}$ *given by*

$$\begin{aligned}\mathfrak{D} &= D(\mathcal{O}) \\ &= \left\{(x+iy, v) \in \mathbf{L} \,\middle|\, 0 \le x+y, x \le 1/2, y \le 1/2, 1 \le x^2+y^2+v^2\right\} \end{aligned} \tag{2.51}$$

is a fundamental domain of $\mathbf{\Lambda}$ *on* $\mathbf{L}$, *i.e., it meets each of the* $\mathbf{\Lambda}$*-orbits and has no distinct inner points belonging to the same orbit.*

The fundamental domain has finite volume with respect to the invariant volume element (2.44).

Proof. By analogy with the symplectic case, let us call the positive real number $v^2 = h(u)$ the *height* of a point $u = (z, v) \in \mathbf{L}$. If $M = \begin{pmatrix} A & B \\ C & D \end{pmatrix} \in \mathbf{S}$ with $g(M) = \begin{pmatrix} \alpha & \beta \\ \gamma & \delta \end{pmatrix}$ and $Z \in \mathbf{H}$ with $u(Z) = (z, v)$, then, by (1.24), (2.42), (2.40), and (2.34), the heights of the points $M\langle Z\rangle$ and $g(u)$ satisfy the relation

$$h(M\langle Z\rangle) = |\det(CZ+D)|^{-2}h(Z) = \left(\frac{v}{\Delta(g,u)}\right)^2 = h(g(u)).$$

It follows that the values of the height of points of the orbit $\mathbf{\Lambda}(u)$ are among the values of the heights of points of the orbit $\Gamma^2\langle Z\rangle$, where $u(Z) = u$. Hence by Lemma 1.15, we conclude that each orbit of $\mathbf{\Lambda}$ on $\mathbf{L}$ contains points u of maximal height and the points can be characterized by inequalities

$$\Delta(g,u) \ge 1 \quad \text{for every } g \in \mathbf{\Lambda}. \tag{2.52}$$

Similarly to the proof of Theorem 1.16, in order to construct a fundamental domain for $\mathbf{\Lambda}$ on $\mathbf{L}$, first of all, we shall choose on the $\mathbf{\Lambda}$-orbit of a point $u'' \in \mathbf{L}$ a point $u' = (z', v')$ of maximal hight. Further, the matrices $g = \begin{pmatrix} \alpha & \beta \\ 0 & \alpha^{-1} \end{pmatrix}$ with $\alpha, \alpha^{-1}, \beta \in \mathcal{O}$ belong to $\mathbf{\Lambda}$, and the corresponding transformations do not change the height:

$$g(u') = g((z', v')) = (\alpha\bar{\alpha}^{-1}z' + \beta\bar{\alpha}^{-1}, v') = (\pm z + \beta', v')$$

with $\beta' = \beta\bar{\alpha}^{-1} \in \mathcal{O}$, since the inclusions $\alpha, \alpha^{-1} \in \mathcal{O}$ mean that α is a unit of $\mathcal{O}$, i.e., $\alpha = \pm 1, \pm i$. It is clear that the sign and the number $\beta' \in \mathcal{O}$ can be chosen such that the point $u = (x+iy, v) = (\pm z + \beta', v')$ satisfies the conditions $0 \le x+y, x \le 1/2, y \le 1/2$. Since the height of u is the same as that of u' and so is again maximal for the considered orbit, the inequality (2.52) for $g = \begin{pmatrix} 0 & 1 \\ -1 & 0 \end{pmatrix}$ shows that

$$\Delta(g,u) = |z|^2 + v^2 = x^2 + y^2 + v^2 \geq 1.$$

Thus, $u \in \mathfrak{D}$.

Suppose now that $u' = \sigma(u)$ for two points $u = (z,v)$, $u' = (z',v') \in \mathfrak{D}$ and $g = \left(\begin{smallmatrix}\alpha & \beta \\ \gamma & \delta\end{smallmatrix}\right) \in \mathbf{A}$. Then by the above, the points u, u' have the same height $v' = v$, whence $\Delta(g,u) = 1$. Similarly, $\Delta(g^{-1},u') = 1$. If $\gamma \neq 0$, the equations are both nontrivial, and so points u, u' belong to the boundary of $\mathfrak{D}$. But if $\gamma = 0$, then, as we have seen above, $z' = \pm z + \beta'$. This implies that $z' = z$, unless the complex numbers $z = x + iy$ and $z' = x' + iy'$ belong to the boundary of the triangle $0 \leq x + y$, $x \leq 1/2$, $y \leq 1/2$. Finiteness of the volume is obvious. □

Exercise 2.11. Let $A = \left(\begin{smallmatrix}2a & b \\ b & 2c\end{smallmatrix}\right)$ be a positive definite even matrix. Check that the matrix

$$\tau = \tau(A) = \frac{1}{\sqrt{2c\sqrt{\det A}}}\begin{pmatrix}2c & -b \\ 0 & \sqrt{\det A}\end{pmatrix} \in \mathrm{SL}_2(\mathbb{R})$$

satisfies ${}^t\tau A\tau = \sqrt{\det A}E$. Assuming that $\gcd(a,b,c) = 1$, show then that the corresponding mapping $g_\tau = g_\tau(A)$ of Exercise 2.8 maps the group

$$\Gamma(A) = S(A) \bigcap \Gamma^2$$

isomorphically onto the group

$$\Gamma(\mathcal{O}(A), \mathfrak{A}(A)) = \left\{ \begin{pmatrix}\alpha & \beta \\ \gamma & \delta\end{pmatrix} \in \mathbf{G} \;\middle|\; \alpha, \delta \in \mathcal{O}(A), \beta \in \mathfrak{A}(A), \gamma \in \mathfrak{A}(A)^{-1} \right\},$$

where $\mathcal{O}(A)$ is the subring of the discriminant $-\det A$ of the ring of integers of the imaginary quadratic field $K = \mathbb{Q}(\sqrt{-\det A})$, $\mathfrak{A}(A) = \frac{c}{\omega}\{a, \omega\}^2$, and $\{a, \omega\}$ is the module in K of the form $a\mathbb{Z} + \omega\mathbb{Z}$.

[Hint: Check first that

$$\Gamma(A) = \left\{ \begin{pmatrix}A & B \\ C & D\end{pmatrix} \;\middle|\; A \in Y_0(\hat{A}), \right.$$

$$\left. B \in X_0(A), C \in X_0(\hat{A}), D \in Y_0(A), A^t D - B^t C = E \right\},$$

where $X_0(A) = X(A) \cap \mathbb{Z}_2^2$, $X(A)$ is the set (2.12), $Y_0(A) = Y(A) \cap \mathbb{Z}_2^2$, $Y(A) = \{Y \in \mathbb{R}_2^2 | YA^tY = \det Y \cdot A, \det Y \geq 0\}$, and $\hat{A} = \det A \cdot A^{-1}$.]

2.4 Radial Series and Eisenstein Series

In this section we shall show that the integral in the representation (2.26) of the radial series $R(s; F, A)$ for the matrix $A = 2E$ can be interpreted as an integral convolution of the restriction of F on $\mathbf{H}$ considered via (2.39) as a function on $\mathbf{L}$ with an Eisenstein series for the discrete subgroup $\mathbf{\Lambda} = \mathrm{SL}_2(\mathbb{Z}[\sqrt{-1}]) \subset \mathbf{G}$.

Let F be a cusp form of the space $\mathfrak{M}_k^2 = \mathfrak{M}_k^2(1, 1)$ and $\widetilde{F}$ the corresponding function (2.47) on the Lobachevsky half-space $\mathbf{L}$. By (2.49), the function $\widetilde{F}$ satisfies the functional equations

$$\Delta(g, u)^{-k}\widetilde{F}(g(u)) = \widetilde{F}(u) \qquad (u \in L, g \in \mathbf{\Lambda}). \tag{2.53}$$

The identity (2.26) can be rewritten in the form

$$(4\pi)^{-s}\Gamma(s)R(s; F, 2E) = \int_{\mathbf{P}} \widetilde{F}(u)v^{s+2}du, \tag{2.54}$$

where

$$\mathbf{P} = \{u = (x + iy, v) \in \mathbf{L} \mid |x| \le 1/2, |y| \le 1/2\}$$

and du is the invariant element of volume (2.44). Note first of all that the set $\mathbf{P}$ can be interpreted as a fundamental domain for the group

$$\mathbf{\Lambda}_0 = \left\{ \begin{pmatrix} \pm 1 & \beta \\ 0 & \pm 1 \end{pmatrix} \in \mathbf{\Lambda} \right\}$$

acting on $\mathbf{L}$. Let

$$\mathbf{\Lambda} = \bigcup_j \mathbf{\Lambda}_0 g_j$$

be a decomposition into disjoint left cosets, and let $\mathcal{D}$ be a fundamental domain of the group $\mathbf{\Lambda}$ on $\mathbf{L}$. Then the set

$$\mathbf{P}' = \bigcup_j g_j(\mathcal{D})$$

can also be taken as a fundamental domain of the group $\mathbf{\Lambda}_0$. Since the integrand in (2.54) is invariant under every transformation of $\mathbf{\Lambda}_0$, it follows that the integral does not change when the domain P is replaced by another sufficiently good fundamental domain of $\mathbf{\Lambda}_0$, for example by $\mathbf{P}'$, provided that the integral converges absolutely. Then we obtain

$$\int_{\mathbf{P}} \widetilde{F}(u)v^{s+2}du = \int_{\mathbf{P}'} \widetilde{F}(u)v^{s+2}du = \sum_j \int_{g_j(\mathcal{D})} \widetilde{F}(u)v^{s+2}du.$$

In each of the last integrals we make the change of variables $u \to g_j^{-1}\langle u\rangle$. Then, by invariance of the volume element du and (2.53), we get

$$\begin{aligned} \int_{\mathbf{P}} \widetilde{F}(u)v^{s+2}du &= \sum_j \int_{\mathcal{D}} \widetilde{F}(u)\frac{v^{s+2}}{\Delta(g_j, u)^{s-k+2}}du \\ &= \int_{\mathcal{D}} \widetilde{F}(u)E^*(u, s-k+2)v^k du, \end{aligned}$$

where

$$E^*(u,s) = v^s \sum_{g\in\mathbf{\Lambda}_0\backslash\mathbf{\Lambda}} \frac{1}{\Delta(g,u)^s}.$$

We note now that two matrices $\begin{pmatrix}\alpha & \beta\\ \gamma & \delta\end{pmatrix}$ and $\begin{pmatrix}\alpha' & \beta'\\ \gamma' & \delta'\end{pmatrix}$ of $\mathbf{\Lambda}$ belong to the same left coset modulo $\mathbf{\Lambda}_0$ if and only if their second rows satisfy $\{\gamma', \delta'\} = \pm\{\gamma, \delta\}$. Moreover, since the Euclidean algorithm holds in the ring of Gaussian integers, for every pair of relatively prime numbers $\gamma, \delta \in \mathcal{O}$ one can find numbers $\alpha, \beta \in \mathcal{O}$ such that $\alpha\delta - \gamma\beta = 1$. It follows that

$$E^*(u,s) = \frac{v^s}{2} \sum_{\substack{\gamma,\delta\in\mathcal{O}\\ \gamma\mathcal{O}+\delta\mathcal{O}=\mathcal{O}}} \frac{1}{(|\gamma z+\delta|^2 + |\gamma|^2 v^2)^s}. \tag{2.55}$$

The series (2.55) is a so-called *Eisenstein series* for the group $\mathbf{\Lambda}$.

Lemma 2.12. *The Eisenstein series* (2.55) *converges absolutely and uniformly on compact subsets of* $\mathbf{L}$ *in each half-plane* $\Re s > 2+\varepsilon$ *with* $\varepsilon > 0$ *and satisfies the relations*

$$E^*(g(u),s) = E^*(u,s) \qquad \text{for all } g \in \mathbf{\Lambda}. \tag{2.56}$$

Proof. For $u = (z, v)$ in a compact subset of $\mathbf{L}$, we clearly have

$$|\gamma z+\delta|^2 + |\gamma|^2 v^2 > c(|\gamma|^2 + |\delta|^2), \tag{2.57}$$

where c is a positive constant. It follows that the series for $2c^{-1}v^{-s}E^*(u,s)$ in the half-plane $\Re s > 2+\varepsilon$ is majorized termwise by the convergent series

$$\sum_{(\gamma,\delta)\neq(0,0)} \frac{1}{(|\gamma|^2+|\delta|^2)^{2+\varepsilon}} = \sum_{n=1}^{\infty} \frac{r_4(n)}{n^{2+\varepsilon}},$$

where $r_4(n)$ is the number of integral solutions of the equation $x_1^2+x_2^2+x_3^2+x_4^2 = n$ (for convergence, see the Jacobi formula for $r_4(n)$).

By (2.34) and (2.43), we obtain

$$\begin{aligned} E^*(g(u),s) &= (v(g(u))^s \sum_{g'\in\mathbf{\Lambda}_0\backslash\mathbf{\Lambda}} \frac{1}{\Delta(g', g(u))^s} \\ &= v^s \Delta(g,u)^{-s} \sum_{g'\in\mathbf{\Lambda}_0\backslash\mathbf{\Lambda}} \frac{\Delta(g,u)^s}{\Delta(g'g,u)^s} = E^*(u,s). \end{aligned}$$

□

We summarize the results of the above considerations in the following lemma.

Lemma 2.13. *For every cusp form* $F \in \mathfrak{N}_k^2$ *with Fourier expansion* (2.4), *the radial Dirichlet series*

$$R_F(s) = \sum_{a=1}^{\infty} \frac{f(2mE)}{m^s}$$

satisfies in a right half-plane of the variable s the identity

$$(4\pi)^{-s}\Gamma(s)R_F(s) = \int_{\mathcal{D}} \widetilde{F}(u)E^*(u, s-k+2)v^k du, \tag{2.58}$$

where $\widetilde{F}(u)$ *is the restriction of* F *on* $\mathbf{H}$ *considered as a function on* $\mathbf{L}$, $E^*(u, s)$ *the Eisenstein series* (2.55), du *the invariant measure* (2.44), *and where* $\mathcal{D}$ *is a fundamental domain for the group* $\mathbf{\Lambda}$ *on* $\mathbf{L}$.

Exercise 2.14. Let $A = \begin{pmatrix} 2a & b \\ b & 2c \end{pmatrix}$ be a positive definite even matrix satisfying $\gcd(a,b,c) = 1$. In the notation of Exercises 2.8 and 2.11, let $\widetilde{F}_A(u) = F(h_\tau^{-1}(u))$ be the function on the space $\mathbf{L}$ corresponding via the map $h_\tau : H(A) \mapsto \mathbf{L}$ to the restriction on $H(A)$ of a cusp form $F \in \mathfrak{N}_k^2$. Assuming that there exists a fundamental domain $D(A)$ of finite volume for the group $\Gamma(\mathcal{O}(A), \mathfrak{A}(A))$ on $\mathbf{L}$, show that the radial Dirichlet series $R(s; F, A)$ satisfies in a right half-plane of the variable s the identity

$$\frac{1}{2}(2\pi)^{-s}(\det A)^{\frac{1-s}{2}}\Gamma(s)R(s; F, A) = \int_{D(A)} \widetilde{F}_A(u)E_A^*(u, s-k+2)v^k du,$$

where

$$E_A^*(u, s) = v^s \sum_{\sigma\in\Gamma_0(\mathcal{O}(A),\mathfrak{A}(A))\backslash\Gamma(\mathcal{O}(A),\mathfrak{A}(A))} \frac{1}{\Delta(\sigma, u)^s},$$

and

$$\Gamma_0(\mathcal{O}(A), \mathfrak{A}(A)) = \left\{ \begin{pmatrix} * & * \\ 0 & * \end{pmatrix} \in \Gamma(\mathcal{O}(A), \mathfrak{A}(A)) \right\}.$$

2.5 Properties of Radial Series for Sums of Two Squares

In this section, in order to apply the integral representation (2.58), we represent the Eisenstein series $E^*(u, s)$ by means of suitable theta series of a positive definite quadratic form in four variables, recall standard properties of the theta series, and finally, prove that the radial Dirichlet series $R(s; F) = R(s; F, 2E)$ has a meromorphic analytic continuation over the whole s-plane and satisfies a functional equation with two gamma–factors.

Eisenstein Series and Theta Series. Let us introduce the theta series

$$\Theta(t, u) = \sum_{\gamma,\delta\in\mathcal{O}} \exp\left(-\frac{\pi t}{v}(|\gamma z+\delta|^2 + |\gamma|^2 v^2)\right) \quad (t > 0, u = (z, v) \in \mathbf{L}).$$

If $t > \eta > 0$ and u is in a compact subset of $\mathbf{L}$, then $v < b$, and by (2.57), the theta series is majorized termwise by the series

$$\sum_{\gamma,\delta\in\mathcal{O}} \exp\left(-\frac{\pi\eta c}{b}(|\gamma|^2 + |\delta|^2)\right) = \sum_{n=-\infty}^{\infty} \exp\left(-\frac{\pi\eta c}{b}n^2\right)^4,$$

and so it converges absolutely and uniformly. If $\Re s$ is sufficiently big, then by using the Euler integral (2.8), we get

$$\int_0^\infty t^{s-1}(\Theta(t,u)-1)dt = \Gamma(s)\left(\frac{v}{\pi}\right)^s \sum_{\gamma,\delta\in\mathcal{O},\,(\gamma,\delta)\neq(0,0)} \frac{1}{(|\gamma z+\delta|^2+|\gamma|^2v^2)^s}.$$

Since each ideal $\mathfrak{A}$ of the ring $\mathcal{O}=\mathbb{Z}[\sqrt{-1}]$ of Gaussian integers is principal, $\mathfrak{A}=\alpha\mathcal{O}$, and its norm $N(\mathfrak{A})$ is $\alpha\bar{\alpha}=|\alpha|^2$, for every nonzero pair $\gamma,\delta\in\mathcal{O}$ with $\gamma\mathcal{O}+\delta\mathcal{O}=\mathfrak{A}$, we have $\gamma=\alpha\gamma'$, $\delta=\alpha\delta'$ with $\gamma'\mathcal{O}+\delta'\mathcal{O}=\mathcal{O}$, hence the last series is equal to

$$\Gamma(s)\left(\frac{v}{\pi}\right)^s \sum_{\mathfrak{A}} \sum_{\substack{\gamma,\delta\in\mathcal{O}\\ \gamma\mathcal{O}+\delta\mathcal{O}=\mathcal{O}}} \frac{1}{N(\mathfrak{a})^s(|\gamma z+\delta|^2+|\gamma|^2v^2)^s} = 2\pi^{-s}\Gamma(s)Z_{\mathcal{O}}(s)E^*(u,s),$$

where $\mathfrak{A}$ ranges through all nonzero integral ideals of the ring $\mathcal{O}$,

$$Z_{\mathcal{O}}(s)=\sum_{\mathfrak{A}}\frac{1}{N(\mathfrak{A})^s} \tag{2.59}$$

is the Dedekind zeta function of the ring $\mathbb{Z}[\sqrt{-1}]$, and $E^*(u,s)$ is the Eisenstein series (2.55). We have proved the identity

$$\int_0^\infty t^{s-1}(\Theta(t,u)-1)dt = 2\pi^{-s}\Gamma(s)Z_{\mathcal{O}}(s)E^*(u,s), \tag{2.60}$$

valid in a right half-plane of the variable s.

Lemma 2.15. (*Inversion formula*) *The theta series $\Theta(t,u)$ satisfies the identity*

$$\Theta(t,u)=\frac{1}{t^2}\Theta\left(\frac{1}{t},u\right) \qquad (t>0, u\in\mathbf{L}).$$

Proof. First we shall prove that for each symmetric positive definite matrix Q of order r, the theta series

$$\theta(t,Q)=\sum_{N\in\mathbb{Z}^r} e^{-\pi t Q[N]} \qquad (Q[N]={}^tNQN) \tag{2.61}$$

satisfies the inversion formula

$$\theta(t,Q)=\frac{1}{\sqrt{t^r\det Q}}\theta\left(\frac{1}{t},Q^{-1}\right). \tag{2.62}$$

Indeed, the function

$$\theta(t,X,Q)=\sum_{N\in\mathbb{Z}^r} e^{-\pi t Q[N+X]} \qquad (X\in\mathbb{R}^r)$$

is a periodic function of X and is equal to its Fourier series

$$\theta(t,X,Q) = \sum_{M\in\mathbb{Z}^r} a(M) e^{2\pi i {}^tNX}, \tag{2.63}$$

where

$$\begin{aligned} a(M) &= \int_0^1 \cdots \int_0^1 \theta(t,X,Q) e^{-2\pi i {}^tMX} dx_1 \cdots dx_r \\ &= \int_{-\infty}^{+\infty} \cdots \int_{-\infty}^{+\infty} e^{-\pi t Q[X] - 2\pi i {}^tMX} dx_1 \cdots dx_r. \end{aligned}$$

Completing the square, we have

$$tQ[X] + 2i{}^tMX = tQ[X + \frac{i}{t}Q^{-1}M] + \frac{1}{t}Q^{-1}[M] \qquad (i=\sqrt{-1}),$$

whence

$$a(M) = e^{-\frac{\pi}{t}Q^{-1}[M]} \int_{\mathbb{R}^r} e^{-\pi t Q[X + \frac{i}{t}Q^{-1}[M]} dX \qquad (dX = dx_1 \cdots dx_r).$$

By Cauchy's theorem, we can write

$$\begin{aligned} a(M) &= e^{-\frac{\pi}{t}Q^{-1}[M]} \int_{\mathbb{R}^r} e^{-\pi t Q[X]} dX \\ &= \frac{1}{\sqrt{t^r \det Q}} e^{-\frac{\pi}{t}Q^{-1}[M]} \int_{\mathbb{R}^r} e^{-\pi {}^tYY} dY = \frac{1}{\sqrt{t^r \det Q}} e^{-\frac{\pi}{t}Q^{-1}[M]}, \end{aligned}$$

where $Y = SX$ with ${}^tSS = tQ$ and $dY = |\det S| dX$, which together with (2.63) for $X = 0$ proves the formula (2.62).

Coming back to the theta function $\Theta(t,u)$, we note that an easy straightforward calculation allows us to interpret it as a special case of the theta function (2.61), namely,

$$\Theta(t,u) = \theta(t,Q), \tag{2.64}$$

where

$$Q = \frac{1}{v}\begin{pmatrix} [u] & 0 & x & y \\ 0 & [u] & -y & x \\ x & -y & 1 & 0 \\ y & x & 0 & 1 \end{pmatrix} \quad \text{if } u = (x+iy, v) \in \mathbf{L} \quad \text{and } [u] = x^2+y^2+v^2.$$

It is easy to check that $\det Q = 1$ and

$$Q^{-1} = \frac{1}{v}\begin{pmatrix} 1 & 0 & -x & -y \\ 0 & 1 & y & -x \\ -x & y & [u] & 0 \\ -y & -x & 0 & [u] \end{pmatrix} = {}^tUQU, U = \begin{pmatrix} 0 & 0 & 0 & 1 \\ 0 & 0 & 1 & 0 \\ 0 & -1 & 0 & 0 \\ -1 & 0 & 0 & 0 \end{pmatrix} \in \mathrm{GL}_4(\mathbb{Z}).$$

Then by (2.64) and (2.62), we obtain

$$\Theta(t,u) = \theta(t,Q) = \frac{1}{t^2}\theta\left(\frac{1}{t}, Q^{-1}\right)$$
$$= \frac{1}{t^2}\theta\left(\frac{1}{t}, Q[U]\right) = \frac{1}{t^2}\theta\left(\frac{1}{t}, Q\right) = \frac{1}{t^2}\Theta\left(\frac{1}{t}, u\right). \square$$

Analytic Continuation and Functional Equation. Now we are finally able to prove the following result.

Proposition 2.16. *Let F be a cusp form of integral weight k for the group Γ^2. Then the function*

$$\Psi(s;F) = (2\pi)^{-2s}\Gamma(s)\Gamma(s-k+2)Z_{\mathcal{O}}(s-k+2)R(s;F),$$

where $\Gamma(s)$ is the gamma function, $Z_{\mathcal{O}}(s)$ is the Dedekind zeta function (2.59) of the ring $\mathcal{O} = \mathbb{Z}[\sqrt{-1}]$, and $R(s;F) = R(s;F,2E)$ is the radial Dirichlet series corresponding to the ray of the matrix $A = 2E = 2\cdot 1_2$ of the sum of two squares, can be continued analytically to the whole s-plane as a meromorphic function having at most two poles at the points $s = k-2$ and $s = k$, and satisfying the functional equation

$$\Psi(2k-2-s;F) = \Psi(s;F).$$

Proof. By (2.58), in a right half-plane of the variable s we have the identity

$$\Psi(s;F) = \pi^{-s}\Gamma(s-k+2)Z_{\mathcal{O}}(s-k+2)\int_{\mathcal{D}}\widetilde{F}(u)E^*(u,s-k+2)v^k du,$$

which, by (2.60), can be rewritten in the form

$$\Psi(s;F) = \frac{1}{2}\pi^{2-k}\int_{\mathcal{D}}\widetilde{F}(u)v^k\left(\int_0^\infty t^{s-k+1}(\Theta(t,u)-1)dt\right)du. \tag{2.65}$$

In view of Lemma 2.15, we obtain

$$\int_0^1 t^{s-k+1}(\Theta(t,u)-1)dt = \int_0^1 t^{s-k+1}\Theta(t,u)dt - \int_0^1 t^{s-k+1}dt$$
$$= \int_0^1 t^{s-k-1}\Theta\left(\frac{1}{t},u\right)dt - \frac{1}{s-k+2}$$
$$= \int_1^\infty t^{k-s-1}\Theta(t,u)dt - \frac{1}{s-k+2}$$
$$= \int_1^\infty t^{k-s-1}(\Theta(t,u)-1)dt + \frac{1}{s-k} - \frac{1}{s-k+2}.$$

Thus, the inner integral in (2.65) can be written in a right half-plane in the form

$$\int_1^\infty t^{s-k+1}(\Theta(t,u)-1)dt + \int_0^1 t^{s-k+1}(\Theta(t,u)-1)dt$$
$$= \int_1^\infty (t^{s-k+1}+t^{k-s-1})(\Theta(t,u)-1)dt + \frac{1}{s-k} - \frac{1}{s-k+2}. \tag{2.66}$$

Substituting this expression in (2.65), we get

$$2\pi^{k-2}\Psi(s;F) = \int_{\mathcal{D}} \widetilde{F}(u)v^k \left(\int_1^\infty (t^{s-k+1}+t^{k-s-1})(\Theta(t,u)-1)dt \right) du$$
$$+ \left(\frac{1}{s-k} - \frac{1}{s-k+2} \right) \int_{\mathcal{D}} \widetilde{F}(u)v^k du. \tag{2.67}$$

Let us take now the domain $\mathcal{D}$ to be the set $\mathfrak{D}$ described in Theorem 2.10, which has a single cusp at the point $v=+\infty$. Since F is a cusp form, the function $\widetilde{F}(u)$ exponentially tends to zero as $u=(z,v)$ tends from within $\mathfrak{D}$ to the cusp. The same is obviously true with respect to the function $\Theta(t,u)-1$ under the inner integral in the last formula. Thus, the integral representation (2.67) gives a meromorphic analytic continuation of $\Psi(s;F)$ to the whole complex plane. This function is regular at all points, except possibly for simple poles at $s=k$ and $s=k-2$, when $\int_{\mathcal{D}} \widetilde{F}(u)v^k du \neq 0$. Furthermore, the right-hand side of (2.67) is invariant under the substitution $s \mapsto 2k-2-s$, which proves the functional equation. □

Exercise 2.17. Prove that the function

$$\Phi(s) = \pi^{-s}\Gamma(s)Z_{\mathcal{O}}(s)E^*(u,s),$$

where $\Gamma(s)$ is the gamma function, $Z_{\mathcal{O}}(s)$ is the zeta function (2.59), and $E^*(u,s)$ is the Eisenstein series (2.55), can be continued analytically to the whole s-plane as a meromorphic function having only two simple poles at the points $s=0$ and $s=2$, and satisfying the functional equation

$$\Phi(2-s) = \Phi(s).$$

[Hint: Use the integral representation (2.60) and identity (2.66) with $s+k-2$ in place of s.]

Chapter 3
Hecke–Shimura Rings of Double Cosets

3.1 An Approach to Multiplicativity

Numerous classical formulas for the numbers of integral representations of integers by integral quadratic forms often show remarkable multiplicative features. Thus, for example, Jacobi's formula for the number of representations $r_4(m)$ of an odd integer m as the sum of four of integer squares,

$$r_4(m) = 8\sum_{d|m} d,$$

expresses the number through the multiplicative function $\varsigma(m) = \sum_{d|m} d$ with the rule of multiplication

$$\varsigma(m)\varsigma(n) = \sum_{d|m,n} d\varsigma\left(\frac{mn}{d^2}\right),$$

whereas Ramanujan's formula for the number of representations of integers as sum of 24 integral squares expresses the number as a linear combination of two multiplicative functions. The numbers of integral representations of integers by an integral positive definite quadratic form can be considered as Fourier coefficients of a modular form, the theta series of the quadratic form. In order to reveal multiplicative properties of the numbers of representations, it is reasonable to consider a more general but formulated in more invariant form problem of multiplicative properties of Fourier coefficients of modular forms. Let us go into some detail. One can argue as follows: let, for example,

$$F(z) = \sum_{m=0}^{\infty} f(m)e^{2\pi imz} \in \mathfrak{M}_k = \mathfrak{M}_k(\Gamma)$$

be a modular form of integral weight k for the group $\Gamma = \Gamma^1$. Generally speaking, a "multiplicativity" of the function $m \mapsto f(m)$ should at least imply that there are

A. Andrianov, *Introduction to Siegel Modular Forms and Dirichlet Series*, Universitext, 63
DOI 10.1007/978-0-387-78753-4_3,

regular relations between this function and functions $m \mapsto f(pm)$ for some prime numbers p. The values $f(pm)$ are the Fourier coefficients of the function

$$F_p(z) = \sum_{m=0}^{\infty} f(pm)e^{2\pi imz} = \frac{1}{p}\sum_{b=0}^{p-1} F\left(\frac{z+b}{p}\right) = \sum_{b=0}^{p-1} F|_k \begin{pmatrix} 1 & b \\ 0 & p \end{pmatrix},$$

where $|_kM$ are the Petersson operators (1.29) for genus $n=1$. If the linear operator $F \mapsto F_p$ maps the space $\mathfrak{M}_k$ into itself, then one could hope to find its eigenfunctions in $\mathfrak{M}_k$, i.e, functions satisfying $F_p = \lambda(p)F$. For each such function one gets relations $f(pm) = \lambda(p)f(m)$ for all $m = 1,2,\ldots$, which provide a kind of multiplicativity. Now, the function F_p belongs to $\mathfrak{M}_k$ if and only if $F_p|_k\gamma = F_p$ for all $\gamma \in \Gamma$. Since $F|_k\gamma = F$ if $\gamma \in \Gamma$, and $|_kM_1M_2 = |_kM_1|_kM_2$ for every two real matrices of order two with positive determinant, the inclusion $F_p \in \mathfrak{M}_k$ would be true if the set of left cosets $\left\{\Gamma\left(\begin{smallmatrix} 1 & 0 \\ 0 & p \end{smallmatrix}\right), \Gamma\left(\begin{smallmatrix} 1 & 1 \\ 0 & p \end{smallmatrix}\right), \ldots \Gamma\left(\begin{smallmatrix} 1 & p-1 \\ 0 & p \end{smallmatrix}\right)\right\}$ coincides up to an order with the set $\left\{\Gamma\left(\begin{smallmatrix} 1 & 0 \\ 0 & p \end{smallmatrix}\right)\gamma, \Gamma\left(\begin{smallmatrix} 1 & 1 \\ 0 & p \end{smallmatrix}\right)\gamma, \ldots \Gamma\left(\begin{smallmatrix} 1 & p-1 \\ 0 & p \end{smallmatrix}\right)\gamma\right\}$ for every $\gamma \in \Gamma$. But this is not true, since, for example, the left coset

$$\Gamma\begin{pmatrix} 1 & 0 \\ 0 & p \end{pmatrix}\begin{pmatrix} 0 & 1 \\ -1 & 0 \end{pmatrix} = \Gamma\begin{pmatrix} 0 & 1 \\ -1 & 0 \end{pmatrix}\begin{pmatrix} p & 0 \\ 0 & 1 \end{pmatrix} = \Gamma\begin{pmatrix} p & 0 \\ 0 & 1 \end{pmatrix}$$

differs from any left coset $\Gamma\left(\begin{smallmatrix} 1 & b \\ 0 & p \end{smallmatrix}\right)$. On the other hand, we note that all the left cosets $\Gamma\left(\begin{smallmatrix} 1 & b \\ 0 & p \end{smallmatrix}\right)\gamma = \Gamma\left(\begin{smallmatrix} 1 & 0 \\ 0 & p \end{smallmatrix}\right)\left(\begin{smallmatrix} 1 & b \\ 0 & 1 \end{smallmatrix}\right)\gamma$ with $\gamma \in \Gamma$ belong to the double coset $\Gamma\left(\begin{smallmatrix} 1 & 0 \\ 0 & p \end{smallmatrix}\right)\Gamma$, and we may replace the sum F_p by the sum over *all* representatives of left cosets contained in that double coset. Thus, in place of the operator $F \mapsto F_p$, we arrive at the famous Hecke operator

$$T(p) : F \mapsto F|T(p) = \sum_{\gamma \in \Gamma\backslash\Gamma\left(\begin{smallmatrix} 1 & 0 \\ 0 & p \end{smallmatrix}\right)\Gamma} F|_kM.$$

Since every right multiplication on an element of Γ only rearranges left Γ-cosets contained in $\Gamma\left(\begin{smallmatrix} 1 & 0 \\ 0 & p \end{smallmatrix}\right)\Gamma$, it easily follows that $T(p)$ maps the space $\mathfrak{M}_k$ into itself. Since p is a prime number, one can easily check that the set

$$\left\{\begin{pmatrix} 1 & 0 \\ 0 & p \end{pmatrix}, \begin{pmatrix} 1 & 0 \\ 1 & p \end{pmatrix}, \ldots, \begin{pmatrix} 1 & p-1 \\ 0 & p \end{pmatrix}, \begin{pmatrix} p & 0 \\ 0 & 1 \end{pmatrix}\right\}$$

is a set of representatives of all different left Γ-classes contained in this double coset. Thus, we conclude that

$$F|T(p) = F_p + F|k\begin{pmatrix} p & 0 \\ 0 & 1 \end{pmatrix} = \sum_{m=0}^{\infty} f(pm)e^{2\pi imz} + p^{k-1}\sum_{m=0}^{\infty} f(m)e^{2\pi ipmz}$$

$$= \sum_{m=0}^{\infty}\left(f(pm) + p^{k-1}c\left(\frac{m}{p}\right)\right)e^{2\pi imz},$$

where $f(\frac{m}{p}) = 0$ if $p \nmid m$. If $F \in \mathfrak{M}_k$ is an eigenfunction for the operator $T(p)$, $F|T(p) = \lambda(p)F$, then, by comparing Fourier coefficients, we have

$$\left(f(pm) + p^{k-1} f\left(\frac{m}{p}\right)\right) = \lambda(p) f(m) \qquad (m = 0, 1, 2, \ldots),$$

and in particular, $f(p) = \lambda(p) f(1)$. Hence we obtain

$$f(1)\left(f(pm) + p^{k-1} f\left(\frac{m}{p}\right)\right) = f(p) f(m) \qquad (m = 0, 1, 2, \ldots),$$

which is not very bad kind of multiplicativity with respect to the prime number p.

The above consideration points to an approach to the problem of multiplicativity of Fourier coefficients of modular forms for subgroups of modular groups as well as the problem of Euler factorization of Dirichlet series formed by the Fourier coefficients, namely the approach based on the study of linear operators on spaces of modular forms associated to certain double cosets of the corresponding subgroup. The double cosets form associative rings, the rings of double cosets or the Hecke–Shimura rings, which we consider in this chapter first in an abstract situation and then for discrete subgroups of the general linear and symplectic groups. In the next chapter we shall study the linear operators on spaces of modular forms corresponding to the rings of double cosets.

Exercise 3.1.

(1) Show that the subspace of cusp forms $\mathfrak{N}_k(\Gamma) \subset \mathfrak{M}_k(\Gamma)$ is invariant with respect to all of the operators $T(p)$.
(2) Show that the function $\Delta'(z) \in \mathfrak{N}_{12}(\Gamma)$ defined in Exercise 1.36 is an eigenfunction for all operators $T(p)$.
(3) (Ramanujan/Mordell) Let

$$\Delta'(z) = \sum_{m=1}^{\infty} f(m) e^{2\pi i m z}.$$

Show that the corresponding radial Dirichlet series has the Euler factorization

$$\sum_{m=1}^{\infty} \frac{f(m)}{m^s} = f(1) \prod_{p \in \mathbb{P}} \frac{1}{1 - \tau(p) p^{-s} + p^{11-2s}},$$

where $\tau(p)$ are the eigenvalues of $T(p)$.

[Hint for (3): using the multiplicative relations

$$\left(f(pm) + p^{11} f\left(\frac{m}{p}\right)\right) = \tau(p) f(m) \qquad (m = 0, 1, 2, \ldots),$$

prove first that for every prime p and every m not divisible by p, we have the identity

$$(1 - \tau(p)t + p^{11}t^2) \sum_{\delta=0}^{\infty} f(mp^{\delta}) t^{\delta} = f(m).]$$

3.2 Abstract Rings of Double Cosets

Definition of Hecke–Shimura rings. In this section we assign to each pair of a multiplicative semigroup with unit and its subgroup, satisfying a simple finiteness condition, an associative ring constructed of cosets of the semigroup modulo the subgroup and consider their basic properties. The simplest situation in which a ring of cosets can be naturally defined is the case of a group Σ and its normal subgroup Λ. In this case each double coset $\Lambda g\Lambda$ of Σ modulo Λ consists of a single left coset, $\Lambda g\Lambda = \Lambda g$; hence one can define the product of double cosets just by

$$(\Lambda g\Lambda)(\Lambda g'\Lambda) = (\Lambda g)(\Lambda g') = (\Lambda gg') = (\Lambda gg'\Lambda)$$

and then extend the multiplication by linearity on finite formal linear combinations of the double cosets with coefficients, say, in the ring $\mathbb{Z}$. As a result, we arrive at the *group ring of the factor group* $\Lambda\backslash\Sigma$ *over* $\mathbb{Z}$. The definition of Hecke–Shimura rings of double cosets just transfers this approach to the situation in which double cosets are unions of finite numbers of left cosets.

Let Σ be a multiplicative semigroup with identity element and Λ a subgroup of Σ. We shall say that the pair Λ, Σ is l*(eft)-finite* if each double coset $\Lambda g\Lambda \subset \Sigma$ is a finite union of left cosets modulo Λ:

$$\Lambda g\Lambda = \bigcup_{i=1}^{\nu(g)} \Lambda g_i \qquad (\nu(g) = \#(\Lambda\backslash\Lambda g\Lambda)). \tag{3.1}$$

The pair is said to be d*(ouble)-finite* if each of the double cosets is both a finite union of left cosets and a finite union of right cosets modulo Λ. For an l-finite pair Λ, Σ we denote by $\mathcal{L} = L(\Lambda, \Sigma)$ the free $\mathbb{Z}$-module consisting of all formal finite linear combinations with coefficients in $\mathbb{Z}$ of the symbols (Λg) with $g \in \Sigma$, which are in one-to-one correspondence with the left cosets Λg of Σ relative to the group Λ. The semigroup Σ naturally operates on $\mathcal{L}$ by multiplication from the right:

$$\Sigma \ni g : \tau = \sum_i a_i(\Lambda g_i) \mapsto \tau g = \sum_i a_i(\Lambda g_i g).$$

Let us consider the submodule $\mathcal{D}$ of $\mathcal{L}$ consisting of all elements, that are invariant under the right multiplication by elements in Λ:

$$\mathcal{D} = D(\Lambda, \Sigma) = \left\{\tau \in \mathcal{L} \,\middle|\, \tau\lambda = \tau \text{ for all } \lambda \in \Lambda\right\}. \tag{3.2}$$

If

$$\tau = \sum_i a_i(\Lambda g_i) \quad \text{and} \quad \tau' = \sum_j a'_i(\Lambda g'_j)$$

are two elements in $\mathcal{D}$, then the linear combination

$$\tau\tau' = \sum_{i,j} a_i a'_j(\Lambda g_i g'_j) \tag{3.3}$$

is independent of the choice of the representatives $g_i \in \Lambda g_i$ and $g'_i \in \Lambda g'_i$ and again belong to $\mathcal{D}$. To see this, we first note that $\tau\tau'$ is independent of the choice of representatives $g_i \in \Lambda g_i$. Now, if we replace the representatives g'_j by $\lambda_j g'_j$ with $\lambda_j \in \Lambda$, by definition of $\mathcal{D}$, we have

$$\sum_i a_i(\Lambda g_i \lambda_j) = \tau\lambda_j = \tau = \sum_i a_i(\Lambda g_i)$$

for each j. Hence

$$\begin{aligned}\sum_{i,j} a_i a'_j(\Lambda g_i \lambda_j g'_j) &= \sum_j a'_j \left(\sum_i a_i(\Lambda g_i \lambda_j)\right) g'_j \\ &= \sum_j a'_j \left(\sum_i a_i(\Lambda g_i)\right) g'_j = \sum_{i,j} a_i a'_j(\Lambda g_i g'_j),\end{aligned}$$

and so the product is independent of the choice of $g'_j \in \Lambda g'_j$. Finally, if $\lambda \in \Lambda$, we have

$$(\tau\tau')\lambda' = \sum_{i,j} a_i a'_j(\Lambda g_i g'_j \lambda) = \tau \cdot (\tau'\lambda) = \tau\tau',$$

whence $\tau\tau' \in \mathcal{D}$. Since the multiplication (3.3) is clearly bilinear and associative, we see that it defines on $D(\Lambda, \Sigma)$ a structure of an associative ring, the *Hecke–Shimura ring of the pair* Λ, Σ *(over* $\mathbb{Z}$*)*.

Basic Rings. The ring $\mathbb{Z}$ in the definition of the Hecke–Shimura ring of an l-finite pair Λ, Σ can be replaced by an arbitrary associative and commutative ring $\mathbb{A}$ with identity. It leads us to the *Hecke–Shimura ring* $D_{\mathbb{A}}(\Lambda, \Sigma)$ *of the pair* Λ, Σ *over* $\mathbb{A}$. The choice of a basic ring $\mathbb{A}$ is insignificant for a majority of the properties of the Hecke–Shimura rings that we are going to consider. As a rule, we take $\mathbb{A} = \mathbb{Z}$, except for the consideration of spherical representations, when we take $\mathbb{A}$ to be the field $\mathbb{Q}$ of rational numbers, and the study of relations between Hecke operators and the Siegel operator, when it is convenient to take $\mathbb{A} = \mathbb{C}$, the field of complex numbers. If the basic ring is not indicated, it means that $\mathbb{A} = \mathbb{Z}$.

Standard Bases and Multiplication Rules. Let Λ, Σ be an l-finite pair. If $g_1, \ldots, g_\nu$ is a complete set of representatives of different left cosets modulo Λ contained in a double coset $\Lambda g \Lambda \subset \Sigma$, then the set $g_1\lambda, \ldots, g_\nu\lambda$ with any $\lambda \in \Lambda$ is clearly still a complete set of representatives of different left cosets modulo Λ contained in a double coset $\Lambda g \Lambda$. It follows that each linear combination of left cosets of the form

$$\tau(g) = \tau_\Lambda(g) = \sum_{g_i \in \Lambda \backslash \Lambda g \Lambda} (\Lambda g_i) \qquad (g \in \Sigma) \tag{3.4}$$

satisfies $\tau(g)\lambda = \tau(g)$ and so belongs to the ring $\mathcal{D} = D(\Lambda, \Sigma)$. By an easy induction, left to the reader, on the number of distinct left cosets that actually occur in a linear combination contained in $\mathcal{D}$, one gets that each such combination is a linear combination of elements of the form $\tau(g)$ with integral coefficients.

Moreover, the elements $\tau(g)$ corresponding to different double cosets $\Lambda g \Lambda$ are obviously linearly independent over $\mathbb{Z}$. We conclude that the elements $\tau(g)$ corresponding to different double cosets $\Lambda g \Lambda$ form a $\mathbb{Z}$-basis of the ring $\mathcal{D}$. The elements $\tau(g)$ will be sometimes referred as *double cosets (modulo* Λ*)* and the ring $\mathcal{D} = D(\Lambda, \Sigma)$ as the *(Hecke–Shimura) ring of double cosets* or just *dc-ring of the pair* Λ, Σ. Note that the element

$$\tau(e) = (\Lambda e) = (\Lambda), \tag{3.5}$$

where e is the identity of Σ, is the identity of the ring $\mathcal{D}$. The following lemma describes the product of double cosets in terms of double cosets.

Lemma 3.2. *The product of double cosets in a dc-ring* $\mathcal{D} = D(\Lambda, \Sigma)$ *can be computed by the following formulas: if*

$$\tau(g) = \sum_{g_i \in \Lambda \backslash \Lambda g \Lambda} (\Lambda g_i) \quad \text{and} \quad \tau(g') = \sum_{g'_j \in \Lambda \backslash \Lambda g' \Lambda} (\Lambda g'_j)$$

are two elements of the form (3.4) *in* $\mathcal{D}$, *then*

$$\tau(g)\tau(g') = \sum_{\Lambda h \Lambda \subset \Lambda g \Lambda g' \Lambda} c(g, g'; h)\tau(h),$$

where h ranges over a set of representatives for the double cosets modulo Λ contained in $\Lambda g \Lambda g' \Lambda$, and a coefficient $c(g, g'; h)$ can be defined as the number of pairs i, j such that $g_i g'_j \in \Lambda h$; the coefficients $c(g, g'; h)$ can also be written in the form

$$c(g, g'; h) = d(g, g'; h)\nu(g')\nu(h)^{-1},$$

where $d(g, g'; h)$ is the number of representatives g_i satisfying $g_i g' \in \Lambda h \Lambda$, and $\nu(g')$, $\nu(h)$ are the numbers of left cosets in the double cosets $\Lambda g' \Lambda$ and $\Lambda h \Lambda$, respectively.

Proof. By definition, we get

$$\tau(g)\tau(g') = \sum_{g_i \in \Lambda \backslash \Lambda g \Lambda} (\Lambda g_i) \sum_{g'_j \in \Lambda \backslash \Lambda g' \Lambda} (\Lambda g'_j) = \sum_{i,j} (\Lambda g_i g'_j).$$

Since each of the products $g_i g'_j$ is clearly contained in $\Lambda g \Lambda g' \Lambda$, it follows that the product can be written in the form

$$\tau(g)\tau(g') = \sum_{\Lambda h \Lambda \subset \Lambda g \Lambda g' \Lambda} c(g, g'; h)\tau(h) = \sum_{\Lambda h \Lambda \subset \Lambda g \Lambda g' \Lambda} c(g, g'; h) \sum_{h_k \in \Lambda \backslash \Lambda h \Lambda} (\Lambda h_k)$$

with some integral coefficients $c(g, g'; h)$. By comparing the coefficients of (Λh) in the two decompositions of the product into left cosets, we conclude that $c(g, g'; h)$ is equal to the number of pairs i, j such that $\Lambda g_i g'_j = \Lambda h$, i.e. $g_i g'_j \in \Lambda h$. Further, it is easy to see that the number $c(g, g'; h)$ depends only on the double cosets of g, g', and h. Thus, the sum

$$\sum_{\Lambda h_k \subset \Lambda \backslash \Lambda h \Lambda} c(g,g';h) = v(h)c(g,g';h)$$

is equal to the number of pairs i, j such that $g_i g'_j \in \Lambda h \Lambda$. Since each representative g'_j can obviously be taken in the form $g'_j = g'\lambda_j$ with $\lambda_j \in \Lambda$, we conclude that the last number equals the number $d(g,g';h)$ of representatives g_i with $g_i g' \in \Lambda h \Lambda$ multiplied by $v(g')$. □

The following lemma often allows one to simplify the decomposition of double cosets into left cosets.

Lemma 3.3. *Let G be a group, Λ a subgroup of G, and g ∈ G. Let*

$$\Lambda = \bigcup_{\lambda_i \in \Lambda \cap g^{-1}\Lambda g \backslash \Lambda} (\Lambda \cap g^{-1}\Lambda g)\lambda_i \quad (\textit{resp.}, \Lambda = \bigcup_{\lambda'_j \in \Lambda/\Lambda \cap g\Lambda g^{-1}} \lambda'_j(\Lambda \cap g\Lambda g^{-1})) \tag{3.6}$$

be decompositions into different cosets. Then

$$\Lambda g \Lambda = \bigcup_{\lambda_i \in \Lambda \cap g^{-1}\Lambda g \backslash \Lambda} \Lambda g \lambda_i \quad (\textit{resp.}, \Lambda g \Lambda = \bigcup_{\lambda'_j \in \Lambda/\Lambda \cap g\Lambda g^{-1}} \lambda'_j g \Lambda), \tag{3.7}$$

where the cosets are disjoint; in particular,

$$v(g) = \#(\Lambda \backslash \Lambda g \Lambda) = [\Lambda : \Lambda \cap g^{-1}\Lambda g]$$
$$(\textit{resp.}, v'(g) = \#(\Lambda g \Lambda/\Lambda) = [\Lambda : \Lambda \cap g\Lambda g^{-1}]). \tag{3.8}$$

Proof. Let us consider, for example, the case of left cosets. It is clear that the right side of (3.7) is contained in the left side. Let $g' = \lambda g \lambda' \in \Lambda g \Lambda$. By (3.6), the element λ' belongs to a left coset $(\Lambda \cap g^{-1}\Lambda g)\lambda_i$, i.e. $\lambda' = \alpha\lambda_i$, where $\alpha \in \Lambda$ and $g\alpha g^{-1} \in \Lambda$. Then we have

$$g' = \lambda g \alpha \lambda_i = \lambda g \alpha g^{-1} g \lambda_i \in \Lambda g \lambda_i,$$

and so the left side of (3.7) is contained in the right side. Now, if a coset $\Lambda g \lambda_i$ meets $\Lambda g \lambda_j$, then we have $\lambda g \lambda_i = \delta g \lambda_j$ with $\lambda, \delta \in \Lambda$, whence $g^{-1}\delta^{-1}\lambda g \lambda_i = \lambda_j$, and so $(\Lambda \cap g^{-1}\Lambda g)\lambda_i = (\Lambda \cap g^{-1}\Lambda g)\lambda_j$, because the element $g^{-1}\delta^{-1}\lambda g = \lambda_j \lambda_i^{-1}$ is obviously contained in $\Lambda \cap g^{-1}\Lambda g$. □

Left and Right Rings of Double Cosets. Suppose now that Λ, Σ is a d-finite pair, i.e., such that each double coset $\Lambda g \Lambda$ in Σ is both a finite union of left cosets and a finite union of right cosets modulo Λ. In this case, the product of double cosets of Σ modulo Λ can be defined not only by a product of right Λ-invariant linear combinations of left cosets modulo Λ, as above, but also by a similar product of left Λ-invariant linear combinations of right cosets modulo Λ. The natural question is whether we get the same product of the double cosets. The solution is not quite trivial, since a double coset can perfectly well consist of different numbers of left and right costs. Nevertheless, the answer is as simple as "yes", at least if Σ is contained in a group.

Theorem 3.4. *Let Λ, Σ be a d-finite pair contained in a group G. Then the product of any double cosets of Σ modulo Λ defined by multiplication of right Λ-invariant linear combinations of left cosets modulo Λ coincides with their product defined by a similar multiplication of left Λ-invariant linear combinations of right cosets modulo Λ.*

The proof of the theorem is a rather technical one; however we shall outline it, because the result has not only philosophical but quite practical value.

First we have to recall definitions and simple properties of commensurability. We say that two subgroups Γ_1 and Γ_2 of a multiplicative group G are *commensurable* if their intersection is of finite index in both Γ_1 and Γ_2. Then we write $\Gamma_1 \sim \Gamma_2$. The relation of commensurability is clearly reflexive and symmetric. It is not hard to see that the relation is also transitive.

Lemma 3.5. *Let G be a group, and Γ a subgroup of G. Then the commensurator of Γ in G,*

$$\widetilde{\Gamma} = \widetilde{\Gamma}_G = \{g \in G | g^{-1}\Gamma g \sim \Gamma\},$$

is a group.

Proof. If $g \in \widetilde{\Gamma}$, then $\Gamma = g(g^{-1}\Gamma g)g^{-1} \sim g\Gamma g^{-1}$, and so $g^{-1} \in \widetilde{\Gamma}$. Now, if $g_1, g_2 \in \widetilde{\Gamma}$, then $g_1^{-1}\Gamma g_1 \sim \Gamma$, hence $g_2^{-1}g_1^{-1}\Gamma g_1 g_2 \sim g_2^{-1}\Gamma g_2 \sim \Gamma$, and so by transitivity, $(g_1 g_2)^{-1}\Gamma g_1 g_2 \sim \Gamma$. □

Lemma 3.6. *Let G be a group, Γ a subgroup of G, and $\widetilde{\Gamma} = \widetilde{\Gamma}_G$ the commensurator of Γ in G. Then the function*

$$\widetilde{\Gamma} \ni g \mapsto \eta(g) = \frac{\nu(g)}{\nu(g^{-1})} \quad \text{with } \nu(h) = [\Gamma : \Gamma \cap h^{-1}\Gamma h]$$

is a homomorphism of the group $\widetilde{\Gamma}$ into the multiplicative group of nonzero rational numbers, which is trivial on the subgroup Γ.

Proof. Let us denote by $\mathcal{X}$ the set of all subgroups of G that are commensurable with Γ. By transitivity, if $\Gamma_1, \Gamma_2 \in \mathcal{X}$, there exists a group $\Gamma' \in \mathcal{X}$ of finite index both in Γ_1 and Γ_2, for example, $\Gamma_1 \cap \Gamma_2$. We then set

$$\eta(\Gamma_1/\Gamma_2) = [\Gamma_1 : \Gamma'][\Gamma_2 : \Gamma']^{-1}. \tag{3.9}$$

It is easy to see that η is independent of the choice of Γ' and satisfies the relations

$$\eta(\Gamma_1/\Gamma_2)\eta(\Gamma_2/\Gamma_3) = \eta(\Gamma_1/\Gamma_3), \quad \eta(g^{-1}\Gamma_1 g/g^{-1}\Gamma_2 g) = \eta(\Gamma_1/\Gamma_2) \ (g \in \widetilde{\Gamma}). \tag{3.10}$$

For $g \in \widetilde{\Gamma}$ and $\Gamma' \in \mathcal{X}$, we set $\eta'(g) = \eta(\Gamma'/g^{-1}\Gamma' g)$. It is readily verified that $\eta'(g)$ does not depend of the choice of Γ'. Hence, by (3.10), we obtain

$$\begin{aligned}\eta'(g_1 g_2) &= \eta(\Gamma'/(g_1 g_2)^{-1}\Gamma' g_1 g_2) \\ &= \eta(\Gamma'/g_1^{-1}\Gamma' g_1)\eta(g_1^{-1}\Gamma' g_1/(g_2^{-1}g_1^{-1}\Gamma' g_1 g_2)) = \eta'(g_1)\eta'(g_2)\end{aligned}$$

if $g_1, g_2 \in \widetilde{\Gamma}$. On the other hand, if $g \in \widetilde{\Gamma}$, then

$$\begin{aligned}\eta'(g) &= \eta(\Gamma/\Gamma \cap g^{-1}\Gamma g)\eta(\Gamma \cap g^{-1}\Gamma g/g^{-1}\Gamma g)\\ &= \nu(g)\eta(g\Gamma g^{-1} \cap \Gamma/\Gamma) = \nu(g)\nu(g^{-1})^{-1} = \eta(g). \quad \square\end{aligned}$$

Proof of Theorem 3.4. It suffices to consider the products of two double cosets. By Lemmas 3.2 and 3.3, we can write

$$\tau(g)\tau(\acute{g}) = \sum_{\Lambda h\Lambda \in \Lambda g\Lambda\acute{g}\Lambda} d(g,\acute{g};h)[\Lambda : \Lambda \cap \acute{g}^{-1}\Lambda\acute{g}][\Lambda : \Lambda \cap h^{-1}\Lambda h]^{-1}\tau(h),$$

where $d(g,\acute{g};h)$ is the number of representatives $g_i \in \Lambda\backslash\Lambda g\Lambda$ satisfying $g_i\acute{g} \in \Lambda h\Lambda$. On the other hand, by similar reasoning for corresponding formal sums of right cosets, we get

$$\begin{aligned}&\sum_{\widetilde{g}_i \in \Lambda g\Lambda/\Lambda} (\widetilde{g}_i\Lambda) \times \sum_{\widetilde{\acute{g}}_j \in \Lambda g'\Lambda/\Lambda} (\widetilde{\acute{g}}_j\Lambda)\\ &= \sum_{\Lambda h\Lambda \in \Lambda g\Lambda\acute{g}\Lambda} \widetilde{d}(g,\acute{g};h)[\Lambda : \Lambda \cap g\Lambda g^{-1}][\Lambda : \Lambda \cap h\Lambda h^{-1}]^{-1} \sum_{\widetilde{h}_k \in \Lambda h\Lambda/\Lambda} (\widetilde{h}_k\Lambda),\end{aligned}$$

where $\widetilde{d}(g,\acute{g};h)$ is the number of representatives $\widetilde{\acute{g}}_j \in \Lambda\acute{g}\Lambda/\Lambda$ satisfying $g\widetilde{\acute{g}}_j \in \Lambda h\Lambda$. In order to prove the equality of the two products, we have only to show that

$$\begin{aligned}&d(g,\acute{g};h)[\Lambda : \Lambda \cap \acute{g}^{-1}\Lambda\acute{g}][\Lambda : \Lambda \cap h^{-1}\Lambda h]^{-1}\\ &= \widetilde{d}(g,\acute{g};h)[\Lambda : \Lambda \cap g\Lambda g^{-1}][\Lambda : \Lambda \cap h\Lambda h^{-1}]^{-1} \qquad (3.11)\end{aligned}$$

if $h \in \Lambda g\Lambda\acute{g}\Lambda$. By Lemma 3.6 for $\Gamma = \Lambda$, we have

$$\begin{aligned}&[\Lambda : \Lambda \cap h^{-1}\Lambda h][\Lambda : \Lambda \cap h\Lambda h^{-1}]^{-1} = \eta(h) = \eta(g)\eta(\acute{g})\\ &= [\Lambda : \Lambda \cap g^{-1}\Lambda g][\Lambda : \Lambda \cap g\Lambda g^{-1}]^{-1}[\Lambda : \Lambda \cap \acute{g}^{-1}\Lambda\acute{g}][\Lambda : \Lambda \cap \acute{g}\Lambda\acute{g}^{-1}]^{-1}.\end{aligned}$$

Thus, the relation (3.11) is equivalent to

$$d(g,\acute{g};h)[\Lambda : \Lambda \cap \acute{g}\Lambda\acute{g}^{-1}] = \widetilde{d}(g,\acute{g};h)[\Lambda : \Lambda \cap g^{-1}\Lambda g]. \qquad (3.12)$$

By the definition and Lemma 3.3, we see that the number $d(g,\acute{g};h)$ can be written in the form

$$d(g,\acute{g};h) = \#\{\lambda \in \Lambda_{(g)}\backslash\Lambda \,|\, g\lambda\acute{g} \in \Lambda h\Lambda\} \quad \text{with } \Lambda_{(g)} = \Lambda \cap g^{-1}\Lambda g. \qquad (3.13)$$

For $\lambda \in \Lambda$, it is easy to see that the double coset $\Lambda g\lambda\acute{g}\Lambda$ depends only on the coset $\Lambda_{(g)}\lambda\Lambda_{(\acute{g}^{-1})}$, and the relation $\Lambda_{(g)}\lambda\delta = \Lambda_{(g)}\lambda$ with $\delta \in \Lambda_{(\acute{g}^{-1})}$ holds if and only if $\delta \in \lambda^{-1}\Lambda_{(g)}\lambda$. Thus we can rewrite the number (3.13) in the form

$$d(g,\acute{g};h) = \sum_{\substack{\lambda \in \Lambda_{(g)}\backslash\Lambda/\Lambda_{(\acute{g}^{-1})},\\ g\lambda\acute{g} \in \Lambda h\Lambda}} [\Lambda_{(\acute{g}^{-1})} : (\Lambda_{(\acute{g}^{-1})} \cap \lambda^{-1}\Lambda_{(g)}\lambda)].$$

Quite similarly,

$$\widetilde{d}(g,\acute{g};h) = \sum_{\substack{\lambda\in\Lambda_{(g)}\backslash\Lambda/\Lambda_{(\acute{g}^{-1})},\\ g\lambda\acute{g}\in\Lambda h\Lambda}} [\Lambda_{(g)} : (\Lambda_{(g)} \cap \lambda\Lambda_{(\acute{g}^{-1})}\lambda^{-1})].$$

Using the symbol (3.9) and its properties (3.10), we obtain

$$\begin{aligned}\widetilde{d}(g,\acute{g};h) &= \sum_{\substack{\lambda\in\Lambda_{(g)}\backslash\Lambda/\Lambda_{(\acute{g}^{-1})},\\ g\lambda\acute{g}\in\Lambda h\Lambda}} \eta(\Lambda_{(g)}/(\Lambda_{(g)} \cap \lambda\Lambda_{(\acute{g}^{-1})}\lambda^{-1}))\\ &= \sum_{\lambda} \eta(\lambda^{-1}\Lambda_{(g)}\lambda/\lambda^{-1}\Lambda_{(g)}\lambda \cap \Lambda_{(\acute{g}^{-1})})\\ &= \sum_{\lambda} \eta(\lambda^{-1}\Lambda_{(g)}\lambda/\Lambda_{(g)})\eta(\Lambda_{(g)}/\Lambda)\eta(\Lambda/\Lambda_{(\acute{g}^{-1})})\\ &\quad \times \eta(\Lambda_{(\acute{g}^{-1})}/\Lambda_{(\acute{g}^{-1})} \cap \lambda^{-1}\Lambda_{(g)}\lambda)\\ &= \sum_{\lambda} \eta(\lambda)\nu(g)^{-1}\nu(\acute{g}^{-1})\eta(\Lambda_{(\acute{g}^{-1})}/\Lambda_{(\acute{g}^{-1})} \cap \lambda^{-1}\Lambda_{(g)}\lambda)\\ &= \nu(g)^{-1}\nu(\acute{g}^{-1})d(g,\acute{g};h),\end{aligned}$$

where ν is defined by (3.8). □

Extension of Antiautomorphisms. For a semigroup Σ and its subgroup Λ, we say that a mapping $g \mapsto g^*$ of Σ into itself is an *antiautomorphism of the pair* Λ, Σ if it is one-to-one on Σ and on Λ and satisfies

$$(g\acute{g})^* = \acute{g}^* g^* \quad \text{for all } g, \acute{g} \in \Sigma. \tag{3.14}$$

The main application of Theorem 3.4 is the following proposition on the extension of antiautomorphisms of d-finite pairs to the corresponding dc-rings.

Proposition 3.7. *Let Λ, Σ be a d-finite pair contained in a group, and let $g \mapsto g^*$ be an antiautomorphism of the second order of this pair. Then the linear mapping of the dc-ring $\mathcal{D} = D(\Lambda, \Sigma)$ into itself defined on the double cosets* (3.4) *by*

$$\tau(g) \mapsto \tau(g)^* = \tau(g^*) \qquad (g \in \Sigma) \tag{3.15}$$

is an antiautomorphism of the second order of the ring $\mathcal{D}$. If, in addition, the mapping is identical on all of the double cosets, then the ring $\mathcal{D}$ is commutative.

Proof. For the first statement it is sufficient to prove that

$$(\tau(g)\tau(\acute{g}))^* = \tau(\acute{g}^*)\tau(g^*) \quad (g, \acute{g} \in \Sigma). \tag{3.16}$$

By Lemma 3.2, we can write the left hand side of (3.16) in the form

$$\sum_{\Lambda h\Lambda \subset \Lambda g\Lambda\acute{g}\Lambda} c(g,\acute{g};h)\tau(h)^* = \sum_{\Lambda h\Lambda \subset \Lambda g\Lambda\acute{g}\Lambda} c(g,\acute{g};h)\tau(h^*),$$

where
$$c(g,\acute{g};h) = \sum_{\substack{g_i \in \Lambda\backslash\Lambda g\Lambda, \acute{g}_j \in \Lambda\backslash\Lambda\acute{g}\Lambda,\\ g_i\acute{g}_j \in \Lambda h}} 1.$$

By Theorem 3.6, we can compute the right-hand side of (3.16) by considering the double cosets $\tau(\phi(g'))$ and $\tau(\phi(g))$ as sums of the right cosets modulo Λ. By similar reasoning, since the map $g \mapsto g^*$ is an antiautomorphism of the second order carrying Λ onto itself, we obtain

$$\sum_{\acute{g}_j^* \in \Lambda\acute{g}^*\Lambda/\Lambda} (\acute{g}_j^*\Lambda) \sum_{g_i^* \in \Lambda g^*\Lambda/\Lambda} (g_i^*\Lambda) = \sum_{\Lambda h^*\Lambda \subset \Lambda\acute{g}^*\Lambda g^*\Lambda} c'(\acute{g}^*, g^*; h^*) \sum_{h_k^* \in \Lambda h^*\Lambda/\Lambda} (h_k^*\Lambda),$$

where
$$c'(\acute{g}^*, g^*; h^*) = \sum_{\substack{\acute{g}_j^* \in \Lambda\acute{g}^*\Lambda/\Lambda; g_i^* \in \Lambda g^*\Lambda/\Lambda;\\ \acute{g}_j^* g_i^* \in h^*\Lambda}} 1.$$

Furthermore, the conditions

$$\acute{g}_j^* \in \Lambda\acute{g}^*\Lambda/\Lambda,\ g_i^* \in \Lambda g^*\Lambda/\Lambda \ \text{ and } \ \Lambda h^*\Lambda \subset \Lambda\acute{g}^*\Lambda g^*\Lambda,\ h_k^* \in \Lambda h^*\Lambda/\Lambda,\ \acute{g}_j^* g_i^* \in h^*\Lambda$$

are equivalent to the conditions

$$\acute{g}_j \in \Lambda\backslash\Lambda\acute{g}\Lambda,\ g_i \in \Lambda\backslash\Lambda g\Lambda \quad \text{and} \quad \Lambda h\Lambda \subset \Lambda g\Lambda\acute{g}\Lambda,\ h_k \in \Lambda\backslash\Lambda h\Lambda,\ g_i\acute{g}_j \in \Lambda h,$$

respectively. It follows that $c'(\acute{g}^*, g^*; h^*) = c(g, \acute{g}; h)$, which proves (3.16).

As to the second statement, the assumption implies that $\tau^* = \tau$ for all $\tau \in \mathcal{D}$; hence we have

$$\tau\acute{\tau} = (\tau\acute{\tau})^* = \acute{\tau}^*\tau^* = \acute{\tau}\tau \quad \text{for all } \tau, \acute{\tau} \in \mathcal{D}. \quad \square$$

Extension of Embeddings. Next we consider embeddings of Hecke–Shimura rings based on inclusions of corresponding pairs. The embeddings turn out to be useful when initial rings seem to be too narrow and we are looking for suitable extensions.

Suppose that Λ, Σ and Λ_0, Σ_0 are two l-finite pairs contained in the same group and satisfying the following conditions:

$$\Lambda_0 \subset \Lambda, \quad \Sigma \subset \Lambda\Sigma_0, \quad \text{and} \quad \Lambda \cap \Sigma_0\Sigma_0^{-1} \subset \Lambda_0, \tag{3.17}$$

where $\Sigma_0^{-1} = \{g^{-1} | g \in \Sigma_0\}$. Since $\Sigma \subset \Lambda\Sigma_0$, each linear combination of left cosets $\tau \in L(\Lambda, \Sigma)$ can be written in the form

$$\tau = \sum_i a_i(\Lambda g_i) \quad \text{with } g_i \in \Sigma_0.$$

Let us consider then the formal linear combination of left cosets

$$\iota(\tau) = \sum_i a_i(\Lambda_0 g_i).$$

The element $\iota(\tau)$ is independent of the choice of representatives $g_i \in \Sigma_0$ in Λg_i, since if $g_i' \in \Lambda g_i \cap \Sigma_0$ are other representatives, the last condition of (3.17) implies that $\lambda_i = g_i' g_i^{-1} \in \Lambda \cap \Sigma_0 \Sigma_0^{-1} \subset \Lambda_0$, whence

$$\iota(\tau)' = \sum_i a_i(\Lambda_0 g_i') = \sum_i a_i(\Lambda_0 \lambda_i g_i) = \sum_i a_i(\Lambda_0 g_i) = \iota(\tau).$$

The condition $\Lambda_0 \subset \Lambda$ implies that the map ι carries different left cosets modulo Λ to different left cosets modulo Λ_0. Thus we obtain an additive embedding of the modules of left cosets $L(\Lambda, \Sigma) \mapsto L(\Lambda_0, \Sigma_0)$.

Proposition 3.8. *Let Λ, Σ and Λ_0, Σ_0 be two l-finite pairs contained in a group and satisfying conditions* (3.17). *Then the restriction of the map ι to the Hecke–Shimura ring $D(\Lambda, \Sigma) \subset L(\Lambda, \Sigma)$ is a ring monomorphism of $D(\Lambda, \Sigma)$ into the ring $D(\Lambda_0, \Sigma_0)$:*

$$\iota : D(\Lambda, \Sigma) \mapsto D(\Lambda_0, \Sigma_0). \tag{3.18}$$

If, moreover,

$$\Sigma_0 \subset \Sigma \quad \textit{and} \quad \nu_\Lambda(g) = \nu_{\Lambda_0}(g) \quad \textit{for all } g \in \Sigma_0, \tag{3.19}$$

where ν stands for the indices (3.8), *then the mapping* (3.18) *is an isomorphism of the rings.*

Proof. The first assertion follows directly from definitions and the inclusion $\Lambda_0 \subset \Lambda$. As to the second, it suffices to show that $\iota(\tau_\Lambda(g)) = \tau_{\Lambda_0}(g)$ for all $g \in \Sigma_0$ and double cosets (3.4) if the conditions (3.19) are fulfilled. Let $\lambda_1, \ldots, \lambda_\nu$ with $\nu = \nu_{\Lambda_0}(g)$ be a set of representatives of the left cosets of Λ_0 modulo $\Lambda_0 \cap g^{-1}\Lambda_0 g$. By Lemma 3.3, we have

$$\tau_{\Lambda_0}(g) = \sum_{i=1}^{\nu} (\Lambda_0 g \lambda_i).$$

On the other hand, the elements $g\lambda_1, \ldots, g\lambda_\nu$ belong to $\Lambda g \Lambda$ and are contained in different left cosets modulo Λ, since a relation $g\lambda_i = \delta\lambda_j$ with $\delta \in \Lambda$ would imply that $\delta = (g\lambda_i\lambda_j^{-1})g^{-1} \subset \Lambda \cap \Sigma_0\Sigma_0^{-1} \subset \Lambda_0$. By (3.19) and (3.7), we obtain

$$\tau_\Lambda(g) = \sum_{i=1}^{\nu} (\Lambda g \lambda_i),$$

and so $\iota(\tau_\Lambda(g)) = \tau_{\Lambda_0}(g)$. □

Exercise 3.9. Let Λ, Σ and Λ_0, Σ_0 be two l-finite pairs contained in a group and satisfying conditions (3.17). Let $g \mapsto g^*$ be an antiautomorphism of the second order of the both pairs. Show then that

$$\iota(\tau)^* = \iota(\tau^*) \quad \text{for all } \tau \in D(\Lambda, \Sigma).$$

Exercise 3.10. Let Λ, Σ be an l-finite pair. Show that the index map

$$D(\Lambda,\Sigma) \ni \sum_i a_i \tau_\Lambda(g_i) \mapsto \nu(\sum_i a_i \tau(g_i)) = \sum_i a_i \nu(g_i),$$

where $\nu(g_i)$ are the indices (3.8), is a ring homomorphism into $\mathbb{Z}$.

3.3 Rings of Double Cosets of the General Linear Group

Although our main objective is the study of dc-rings of the symplectic group, we shall start our consideration of concrete rings with the case of the general linear group because this case is easier, it coincides with the symplectic case for matrices of order two, and, as we shall see later, in some aspects the symplectic case can be reduced to the general linear case.

Global Rings. Our basic pair here will be the pair

$$\Lambda = \Lambda^n = \mathrm{GL}_n(\mathbb{Z}) = \left\{ g \in \mathbb{Z}_n^n \,\middle|\, \det g = \pm 1 \right\}, \quad \Sigma = \Sigma^n = \left\{ g \in \mathbb{Z}_n^n \,\middle|\, \det g \neq 0 \right\}$$

with $n = 1, 2, \ldots$.

Lemma 3.11. *The pair Λ^n, Σ^n is d-finite for every n.*

Proof. By considering the entries of matrices in Λ modulo a positive integer q, we get a homomorphism of Λ into the finite group $\mathrm{GL}_n(\mathbb{Z}/q\mathbb{Z})$. The kernel of this homomorphism is the *principal congruence subgroup of level q of Λ*,

$$\Lambda(q) = \Lambda^n(q) = \left\{ \lambda \in \Lambda \,\middle|\, \lambda \equiv 1_n \pmod{q} \right\}, \tag{3.20}$$

which is therefore a normal subgroup of finite index in Λ.

If $g \in \Lambda$, then clearly the groups $g\Lambda(d)g^{-1}$ and $g^{-1}\Lambda(d)g$, where $d = |\det g|$, are both contained in Λ. It follows that the groups $\Lambda \cap g^{-1}\Lambda g$ and $\Lambda \cap g\Lambda g^{-1}$ both contain the group $\Lambda(d)$, and so have finite indices in Λ. The lemma follows then from Lemma 3.3. □

The lemma allows us to define the Hecke–Shimura ring (3.1),

$$\mathcal{H} = \mathcal{H}^n = D(\Lambda^n, \Sigma^n), \tag{3.21}$$

of the pair Λ^n, Σ^n. As we have seen in the previous section, elements of the form (3.4), which, for the group $\Lambda = \Lambda^n$, will be denoted by

$$t(g) = \tau_\Lambda(g) \qquad (g \in \Sigma^n), \tag{3.22}$$

corresponding bijectively to the double cosets $\Lambda g \Lambda$ contained in Σ, can be taken as a (free) basis of the ring $\mathcal{H}$ (over $\mathbb{Z}$). The following theorem gives a convenient parametrization of the double cosets.

Theorem 3.12. (*The theorem on elementary divisors*). *Each of the double cosets $\Lambda g \Lambda$ with $\Lambda = \Lambda^n$ and $g \in \Sigma = \Sigma^n$ contains a unique representative of the form*

$$\mathrm{ed}(g) = \mathrm{diag}(d_1, \ldots, d_n) \quad \text{with } d_1, \ldots, d_n \in \mathbb{N} \text{ and } d_1 | d_2 | \cdots | d_n. \tag{3.23}$$

First we cite a simple lemma, which like Lemma 1.10 is an easy consequence of the Euclidean algorithm.

Lemma 3.13. *Let $\mathbf{l}$ be a nonzero integral n-row (resp.,n-column) and d the greatest common divisor of the entries of $\mathbf{l}$. Then there exists a matrix $\lambda \in \Lambda^n$ such that $\mathbf{l}\lambda = (d, 0, \ldots, 0)$ (resp., $\lambda \mathbf{l} = {}^t(d, 0, \ldots, 0)$)*.

Proof of Theorem 3.12. First we note that elementary integral operations on rows of matrices $g \in \Sigma$ such as a permutation of rows and addition to one row of another row multiplied by an integer (resp., similar operations on columns) can be derived by multiplication of the matrix from the left (resp., from the right) by a matrix of Λ and so does not change the coset $\Lambda g \Lambda$.

In order to prove that a representative of the form (3.23) exists, we apply induction to n. If $n = 1$, there is nothing to prove. Suppose that $n > 1$, and the theorem has already been proved for matrices of order $n - 1$. Denote by $\delta = \delta(g)$ the minimum of the greatest common divisor of entries of rows of a matrix $g \in \Sigma^n$. By rearranging rows and applying Lemma 3.13, we may replace g by a matrix $g' = (g'_{ij}) \in \Lambda g \Lambda$ with $\delta(g') = \delta(g)$ and the first row $(g'_{11}, \ldots, g'_{1n}) = (\delta, 0, \ldots, 0)$. Now we apply induction to δ. If $\delta = 1$, then by adding to other rows suitable multiples of the first row, we can replace g' by a matrix $g'' = (g''_{ij})$ in the same double coset with $g''_{21} = 0, \ldots, g''_{n1} = 0$, and the theorem is true by the induction assumption on n. But if $\delta > 1$, then by similar reasoning, we can replace g' by a the matrix $g'' = (g''_{ij})$ in the same double coset with $1 \le g''_{21} \le \delta, \ldots, 1 \le g''_{n1} \le \delta$. If the greatest common divisor of entries of each row of the matrix g'' is equal to δ, then $g''_{21} = \cdots = g''_{n1} = \delta$, all entries of g'' are divisible by δ, and we clearly can replace g'' by a matrix of the form $\begin{pmatrix} \delta & 0 \\ 0 & \delta g_1 \end{pmatrix}$ with $g_1 \in \Sigma^{(n-1)}$, which is again contained in $\Lambda g \Lambda$, and we can use the induction assumption on n. Otherwise, $\delta(g'') < \delta$, and we may apply the induction assumption on δ. This proves the existence of a representative of the form (3.23).

If $D = \mathrm{diag}(d_1, \ldots, d_n)$ and $D' = \mathrm{diag}(d'_1, \ldots, d'_n)$ are two matrices of the form (3.23) satisfying $D' = \lambda D \lambda'$ with $\lambda, \lambda' \in \Lambda$, then the Binet–Cauchy formula implies that each r-minor of D' with $1 \le r \le n$ is divisible by $d_1 \cdots d_r$; in particular, $d_1 \cdots d_r | d'_1 \cdots d'_r$. Similarly, $d'_1 \cdots d'_r | d_1 \cdots d_r$. Hence $D = D'$. □

The matrix $\mathrm{ed}(g) = \mathrm{diag}(d_1, \ldots, d_n)$ is called the *matrix of elementary divisors of* g, and the numbers $d_r = d_r(g)$ are the *elementary divisors of* g. The Binet–Cauchy formula implies that the product $d_1 \cdots d_r$ is equal to the greatest common divisor of the minors of order r of g; in particular

$$d_1 \cdots d_n = |\det g|. \tag{3.24}$$

It follows that the additive structure of the ring $\mathcal{H}$ is very simple: it is just a lattice spanned by the generators

$$t[d_1,\ldots,d_n]=t(\mathrm{diag}(d_1,\ldots,d_n)) \quad \text{with } d_1,\ldots,d_n\in\mathbb{N} \text{ and } d_1|d_2|\cdots|d_n, \tag{3.25}$$

where $t(g)$ are the double cosets (3.22). Let us now turn to the multiplicative structure. First of all, we have the following theorem.

Theorem 3.14. *The dc-ring $\mathcal{H}^n=D(\Lambda^n,\Sigma^n)$ is commutative for every n.*

Proof. The mapping $g\mapsto {}^tg$, where ${}^tg$ is the transpose of g, is clearly an antiautomorphism of order two of the pair Λ^n, Σ^n that does not change diagonal matrices. Theorem 3.12 and Proposition 3.7 complete the proof. □

As to explicit rules of multiplication, we have the following basic theorem.

Theorem 3.15. *The double cosets* (3.25) *satisfy the rules*

$$t[d_1,\ldots,d_n]t[d'_1,\ldots,d'_n]=t[d_1d'_1,\ldots,d_nd'_n] \quad \text{if } \gcd(d_n/d_1,d'_n/d'_1)=1. \tag{3.26}$$

In particular,

$$[d]t[d_1,\ldots,d_n]=t[d_1,\ldots,d_n][d]=t[dd_1,\ldots,dd_n], \tag{3.27}$$

where

$$[d]=[d]_n=t(d\cdot 1_n)=(\Lambda d\cdot 1_n). \tag{3.28}$$

Proof. Since the double coset (3.28) consists of a single left coset, the relation (3.27) follows directly from the definition of multiplication in *dc*-rings. It follows from (3.27) that it is sufficient to prove the general relations (3.26) only when $d_1=d'_1=1$ and $\gcd(d_n,d'_n)=1$, or under the more general assumption that the numbers $d_1\cdots d_n$ and $d'_1\cdots d'_n$ are coprime. In order to prove it, we shall use the following lemma.

Lemma 3.16. *The elements*

$$t(a)=t^n(a)=\sum_{\substack{1\le d_1|\cdots|d_n,\\ d_1\cdots d_n=a}} t[d_1,\ldots,d_n]\in\mathcal{H}^n \tag{3.29}$$

with $a\in\mathbb{N}$ have decompositions into left cosets of the form

$$t(a)=\sum_{g\in\Lambda\backslash\Sigma,\,|\det g|=a}(\Lambda g) \tag{3.30}$$

and satisfy the relations

$$t(a)t(a')=t(aa') \qquad \text{if } \gcd(a,a')=1. \tag{3.31}$$

Proof of the lemma. The decompositions (3.30) follow from (3.24) and Theorem 3.12.

Let $g_1,\ldots,g_\nu$ and $g'_1,\ldots,g'_\mu$ be systems of representatives for the left cosets modulo Λ of matrices in Σ with determinants $\pm a$ and $\pm a'$, respectively. We have to show that the product of these systems $g_1g'_1,\ldots,d_\nu d'_\mu$ is a system of representatives for the left cosets modulo Λ of matrices in Σ with determinant $\pm aa'$, provided that a and a' are coprime. First, all of the products have determinant $\pm aa'$. Further, the products belong to different left cosets modulo Λ: if $\lambda g_i g'_j = g_k g'_l$ with $\lambda \in \Lambda$, then the matrix $\gamma = g_k^{-1}\lambda g_i = g'_l(g'_j)^{-1}$ has determinant ± 1 and is integral, since its products by coprime numbers $\det g_k = \pm a$ and $\det g'_j = \pm a'$ are both integral, and hence $\gamma \in \Lambda$ and $\gamma g'_j = g'_l$, so $g'_j = g'_l$ and $\lambda g_i = g_k$, whence $g_i = g_k$. Finally, if $h \in \Sigma$ and $\det h = \pm aa'$, then by Theorem 3.12, it follows easily that h can be written in the form $h = h_1 h_2$ with $h_1, h_2 \in \Sigma$ satisfying $\det h_1 = \pm a$ and $\det h_2 = \pm a'$. We have $h_2 = \delta g'_j$ and $h_1 \delta = \gamma g_i$ with $\delta, \gamma \in \Lambda$; hence $h = h_1 h_2 = \gamma g_i g'_j \in \Lambda g_i g'_j$. □

We can now return to the proof of the theorem. Let $g = \mathrm{ed}(g)$ and $g' = \mathrm{ed}(g')$ be two matrices of elementary divisors satisfying $\det g = a$ and $\det g' = a'$. Since clearly $gg' \in \Lambda g \Lambda g' \Lambda$, it follows from Lemma 3.2 that the product of corresponding double cosets (3.22) has the form

$$t(g)t(g') = t(gg') + s(g,g'),$$

where $s(g,g')$ is a linear combination of double cosets with nonnegative coefficients. Hence, we obtain

$$\begin{aligned} t(a)t(a') &= \left(\sum_{g=\mathrm{ed}(g),\,\det g = a} t(g)\right)\left(\sum_{g'=\mathrm{ed}(g'),\,\det g' = a'} t(g')\right) \\ &= \sum_{\substack{g=\mathrm{ed}(g),\,\det g=a,\\ g'=\mathrm{ed}(g'),\,\det g'=a'}} t(gg') + \sum_{\substack{g=\mathrm{ed}(g),\,\det g=a,\\ g'=\mathrm{ed}(g'),\,\det g'=a'}} s(g,g'). \end{aligned}$$

Since the numbers a and a' are coprime, it follows easily that the first sum on the right is equal to $t(aa')$, which equals $t(a)t(a')$ by Lemma 3.14. Hence, the second sum on the right is zero, and so each of the sums $s(g,g')$ is zero too. □

Exercise 3.17. For $g, g' \in \Sigma^n$ satisfying $\gcd(d_n(g)/d_1(g), d_n(g')/d_1(g')) = 1$, prove that the elementary divisors satisfy the relations

$$d_i(gg') = d_i(g)d_i(g') \quad (i = 1, 2, \ldots, n).$$

The following useful lemma gives a convenient description of representatives for the left cosets of Σ^n modulo Λ^n. A matrix $g = (g_{ij}) \in \Sigma^n$ will be called *reduced* if its entries satisfy

$$\begin{cases} g_{11} > 0, & \\ 0 \le g_{ij} < g_{jj} & \text{if } 1 \le i < j \le n, \\ g_{ij} = 0 & \text{if } i > j. \end{cases} \tag{3.32}$$

Lemma 3.18. *Each left coset $\Lambda^n g \subset \Sigma^n$ contains its unique reduced representative.*

Proof. We shall apply induction to n. If $n = 1$ there is nothing to prove. Let $n > 1$ and suppose that the lemma has already been proved for $n - 1$. By Lemma 3.13 applied to the first column of g, we conclude that the coset Λg contains a representative g' of the form $\begin{pmatrix} d_1 & * \\ 0 & g_1 \end{pmatrix}$ with $d_1 > 0$ and $g_1 \in \Sigma^{n-1}$. By the inductive hypothesis, we can write $g_1 = \lambda g_0$, where $\lambda \in \Lambda^{n-1}$ and g_0 is reduced. It follows from the Euclidean algorithm that there exists a row $\mathbf{l} \in \mathbb{Z}_1^{n-1}$ such that the matrix

$$\begin{pmatrix} 1 & \mathbf{l} \\ 0 & 1_{n-1} \end{pmatrix} \begin{pmatrix} 1 & 0 \\ 0 & \lambda^{-1} \end{pmatrix} g' \in \Lambda g$$

is reduced. The uniqueness follows easily by contradiction. □

Exercise 3.19. Show that for every $l \in \mathbb{N}$, the element $t(l) = t^2(l)$ of $\mathcal{H}^2$ has the following decomposition into left cosets

$$t(l) = \sum_{\substack{a,b \in \mathbb{N},\, ad = l, \\ 0 \le b < d}} \left(\Lambda^2 \begin{pmatrix} a & b \\ 0 & d \end{pmatrix} \right).$$

Using this decomposition, prove that for each prime p and positive integer δ, we have the relation

$$t(p)t(p^\delta) = t(p^{\delta+1}) + p[p]t(p^{\delta-1}),$$

where $[p] = [p]_2$; deduce from these relations that for every l, $l' \in \mathbb{N}$,

$$t(l)t(l') = \sum_{d \mid \gcd(l,l')} d \cdot [d] t(ll'/d^2).$$

Local Subrings. The study of the *global* Hecke–Shimura ring of the general linear group can be reduced to consideration of their *local subrings* corresponding to all prime numbers. For a prime number p we define a subsemigroup Σ_p of Σ by

$$\Sigma_p = \Sigma_p^n = \left\{ g \in \Sigma^n \,\middle|\, \det g | p^\infty \right\}, \tag{3.33}$$

where the notation

$$a | b^\infty \tag{3.34}$$

means that a divides a power of b. The subset of $\mathcal{H}$ consisting of all linear combinations of left or double cosets of matrices whose determinants divide powers of the prime number p,

$$\mathcal{H}_p = \left\{ \sum_i a_i (\Lambda g_i) \in \mathcal{H} \,\middle|\, \det g_i | p^\infty \right\},$$

is clearly a subring called a *local subring* or the *p-subring of* $\mathcal{H}$. The p-subring can be itself interpreted as the Hecke–Shimura ring of the pair $\Lambda = \Lambda^n$, $\Sigma_p = \Sigma_p^n$:

$$\mathcal{H}_p = \mathcal{H}_p^n = D(\Lambda^n, \Sigma_p^n). \tag{3.35}$$

Theorem 3.20. *The ring $\mathcal{H}^n$ is generated by the local subrings $\mathcal{H}_p^n$ for all prime numbers p.*

Proof. If $d_1, d_2, \ldots, d_n$ are elementary divisors of a matrix $g \in \Sigma = \Sigma^n$, for a prime number p, we define the *matrix of elementary p-divisors of g* by

$$\mathrm{ed}_p(g) = \mathrm{diag}\left(p^{\nu_p(d_1)}, p^{\nu_p(d_2)}, \ldots, p^{\nu_p(d_n)}\right), \tag{3.36}$$

where $p^{\nu_p(d_1)}, p^{\nu_p(d_2)}, \ldots, p^{\nu_p(d_n)}$, with $\nu_p(d_i)$ denoting the highest power of p dividing d_i, are the *elementary p-divisors of g*. Note that for a fixed g, we have $\mathrm{ed}_p(g) = 1_n$ for all but a finite number of primes p, and

$$\prod_{p \in \mathbb{P}} \mathrm{ed}_p(g) = \mathrm{ed}(g),$$

where $\mathbb{P}$ denotes the set of all prime numbers. Since $g \in \Lambda \mathrm{ed}(g) \Lambda$, we conclude that each $g \in \Sigma$ can be written in the form

$$g = \prod_{p \in \mathbb{P}} g_p \quad \text{with } \mathrm{ed}(g_p) = \mathrm{ed}_p(g).$$

By Theorem 3.15, we then obtain the decomposition of the double coset (3.22) as

$$t(g) = \prod_{p \in \mathbb{P}} t(g_p), \tag{3.37}$$

where $t(g_p) \in \mathcal{H}_p$, and all but a finite number of $t(g_p)$ are equal to the identity (of the ring $\mathcal{H}$). The theorem follows. □

We now turn to consideration of the structure of the local subrings. A local ring $\mathcal{H}_p = \mathcal{H}_p^n$ consists of integral linear combinations of linearly independent elements of the form

$$t[p^{\delta_1}, \ldots, p^{\delta_n}] \quad \text{with } 0 \le \delta_1 \le \cdots \le \delta_n. \tag{3.38}$$

An element of this form is called *primitive* if $\delta_1 = 0$, and *imprimitive* if $\delta_1 > 0$. An element of $\mathcal{H}_p$ is called *primitive* (resp., *imprimitive*) if it is a linear combination of primitive (resp., imprimitive) elements. It is clear that each element $t \in \mathcal{H}_p$ has a unique decomposition $t = t^{pr} + t^{im}$, where t^{pr} is primitive and t^{im} is imprimitive. It follows from (3.27) that the set $\mathcal{J}$ of all imprimitive elements of $\mathcal{H}_p$ is the principal ideal generated by the element $[p] = [p]_n$, $\mathcal{J} = [p]\mathcal{H}_p$.

Lemma 3.21. *The linear mapping $\phi : \mathcal{H}_p^n \mapsto \mathcal{H}_p^{n-1}$ satisfying*

$$\phi\left(t[p^{\delta_1}, \ldots, p^{\delta_n}]\right) = \begin{cases} t[p^{\delta_2}, \ldots, p^{\delta_n}] & \text{if } 0 = \delta_1 \le \delta_2 \le \cdots \le \delta_n, \\ 0 & \text{if } 0 < \delta_1 \le \delta_2 \le \cdots \le \delta_n, \end{cases}$$

is a ring epimorphism with kernel $\mathcal{J}$.

Proof. It is obvious that this mapping is an epimorphism of $\mathbb{Z}$-modules. It follows from (3.27) that the kernel ϕ equals $\mathcal{J}$. Thus, we have only to check that it is a ring homomorphism. For this, it suffices to verify the relations

$$\phi\left(t[1,p^{\delta_2},\ldots,p^{\delta_n}]t[1,p^{\delta'_2},\ldots,p^{\delta'_n}]\right)=t[p^{\delta_2},\ldots,p^{\delta_n}]t[p^{\delta'_2},\ldots,p^{\delta'_n}], \quad (3.39)$$

where $0\le\delta_2\le\cdots\le\delta_n$ and $0\le\delta'_2\le\cdots\le\delta'_n$. Let us set

$$g=\operatorname{diag}(1,p^{\delta_2},\ldots,p^{\delta_n})=\begin{pmatrix}1&0\\0&g_0\end{pmatrix},\quad g'=\operatorname{diag}(1,p^{\delta'_2},\ldots,p^{\delta'_n})=\begin{pmatrix}1&0\\0&g'_0\end{pmatrix}.$$

By Lemma 3.2 and the definition of ϕ, we obtain

$$\phi(t(g)t(g'))=\sum_h c(g,g';h)t(h_0),$$

where h ranges over the matrices of the form $\left(\begin{smallmatrix}1&0\\0&h_0\end{smallmatrix}\right)$, with $h_0=\operatorname{diag}(p^{\gamma_2},\ldots,p^{\gamma_n})$, $0\le\gamma_2\le\cdots\le\gamma_n$, and $\gamma_2+\cdots+\gamma_n=\delta_2+\cdots+\delta_n+\delta'_2+\cdots+\delta'_n$. Similarly, in the ring $\mathcal{H}_p^{n-1}$ we have the relation

$$t(g_0)t(g'_0)=\sum_{h_0}c(g_0,g'_0;h_0)t(h_0),$$

where h_0 ranges over the matrices indicated above. Thus, in order to prove (3.39), we have only to show that

$$c(g,g';h)=c(g_0,g'_0;h_0). \quad (3.40)$$

Since the coefficient $c(g,g';h)$ depends only on the double cosets of matrices g, g', h modulo Λ, we may replace h by $h'=\left(\begin{smallmatrix}h_0&0\\0&1\end{smallmatrix}\right)$. By Lemmas 3.2 and 3.18, the coefficient $c(g,g';h)$ is equal to the number of pairs d, d' of reduced matrices satisfying the conditions $d\in\Lambda g\Lambda$, $d'\in\Lambda g'\Lambda$, and $dd'=\lambda h'$. Since $\lambda=dd'(h')^{-1}$, the matrix λ is an upper triangular matrix; in particular, it can be written in the form $\lambda=\left(\begin{smallmatrix}\lambda_0&*\\0&\lambda_{nn}\end{smallmatrix}\right)$ with $\lambda_0\in\Lambda^{n-1}$ and $\lambda_{nn}=\pm1$. We can write $d=\left(\begin{smallmatrix}d_0&\mathbf{v}\\0&d_{nn}\end{smallmatrix}\right)$, $d'=\left(\begin{smallmatrix}d'_0&\mathbf{v}'\\0&d'_{nn}\end{smallmatrix}\right)$, where d_0 and d'_0 are reduced matrices of order $n-1$, $d_{nn}>0$, and all of the entries of the columns $\mathbf{v}$ and $\mathbf{v}'$ are nonnegative integers smaller than d_{nn} and d'_{nn}, respectively. Then the relation $dd'=\lambda h'$ implies that $d_{nn}d'_{nn}=\lambda_{nn}$ and $d_0d'_0=\lambda_0h_0$. By the first of the relations we have $d_{nn}=d'_{nn}=1$, and hence $\mathbf{v}=\mathbf{v}'=\mathbf{0}$. Thus, $d=\left(\begin{smallmatrix}d_0&0\\0&1\end{smallmatrix}\right)$, $d'=\left(\begin{smallmatrix}d'_0&0\\0&1\end{smallmatrix}\right)$, where d_0 and d'_0 are reduced matrices of order $n-1$ satisfying $d_0d'_0=\lambda_0h_0$ with $\lambda_0\in\Lambda^{n-1}$. It is clear that $d_0\in\Lambda^{n-1}g_0\Lambda^{n-1}$ and $d'_0\in\Lambda^{n-1}g'_0\Lambda^{n-1}$. Conversely, arbitrary matrices d and d' satisfying these conditions are reduced, belong to the double cosets $\Lambda g\Lambda$ and $\Lambda g'\Lambda$, respectively, and satisfy $dd'=\lambda h'$ with $\lambda\in\Lambda$. This proves the equality (3.40). □

Theorem 3.22. *For every $n=1,2,\ldots$ and every prime number p, the local ring $\mathcal{H}_p^n$ is generated by the elements*

$$\pi_i(p)=\pi_i^n(p)=t(D_i)\quad \text{with } D_i=D_i^n=\begin{pmatrix}1_{n-i}&0\\0&p\cdot 1_i\end{pmatrix}\ (1\le i\le n); \quad (3.41)$$

the elements $\pi_1(p),\ldots,\pi_n(p)$ are algebraically independent over $\mathbb{Z}$, where $\mathbb{Z}$ is identified with the isomorphic subring $\mathbb{Z}\big[[1]_n\big] \subset \mathcal{H}_p^n$.

Proof. For the first assertion it suffices to show that each element t of the form (3.38) is a polynomial in $\pi_1(p),\ldots,\pi_n(p)$ with coefficients in $\mathbb{Z}$. We shall use induction on n and $N = \delta_1 + \cdots + \delta_n$. For $n = 1$ the assertion is clear, since $t[p^{\delta}] = \pi_1^1(p)^{\delta}$. Let us assume that $n > 1$ and the assertion has already been proved for $n-1$. If $N = 1$, then $t = \pi_1(p)$. Let $N > 1$ and the assertion is true for $N' < N$. If $\delta_1 \geq 1$, by (3.27) we have $t = \pi_n(p)^{\delta_1} t'$, where $t' = t[1, p^{\delta_2-\delta_1},\ldots,p^{\delta_2-\delta_1}]$ with $N' = N - n\delta_1 < N$, and hence t' is an integral polynomial in elements (3.41). But if $\delta_1 = 0$, then the element t is primitive, and so by the inductive assumption on n, the image $t'' = t[p^{\delta_2},\ldots,p^{\delta_n}] = \phi(t)$ of t under the mapping of Lemma 3.21 has the form $t'' = P(\pi_1^{n-1}(p),\ldots,\pi_{n-1}^{n-1}(p))$, where

$$P(x_1,\ldots,x_{n-1}) = \sum_{i_1+2i_2+\cdots+(n-1)i_{n-1}=N} a_{i_1,\ldots,i_{n-1}} x_1^{i_1} \cdots x_{n-1}^{i_{n-1}},$$

since each double coset (h) entering into the product $\pi_1^{n-1}(p)^{i_1} \cdots \pi_{n-1}^{n-1}(p)^{i_{n-1}}$ must satisfy $|\det h| = p^{i_1+2i_2+\cdots+(n-1)i_{n-1}} = p^{\delta_2+\cdots+\delta_n} = p^N$. By the definition of the mapping ϕ, we conclude that the element $t_1 = t - P(\pi_1^n(p),\ldots,\pi_{n-1}^n(p))$ belongs to the kernel of ϕ, and hence, by Lemma 3.21, is divisible by $[p] = \pi_n^n(p)$ in the ring $\mathcal{H}_p$: $t_1 = \pi_n^n(p)t_1'$. By the above, t_1 is a linear combination of elements $t[p^{\delta_1'},\ldots,p^{\delta_n'}]$ with $\delta_1' + \cdots + \delta_n' = N$, then t_1' is a linear combination of elements $t[p^{\delta_1''},\ldots,p^{\delta_n''}]$ with $\delta_1'' + \cdots + \delta_n'' = N - n$. The first assertion follows.

Now by induction on n we shall prove that the elements (3.41) are algebraically independent over $\mathbb{Z}$. It is clear if $n = 1$, since the elements $\pi_1^1(p)^{\delta} = t_1(p^{\delta})$ with $\delta = 0,1,2,\ldots$ correspond to distinct double cosets modulo $\Lambda^1 = \{\pm 1\}$ and so are linearly independent. Let us assume now that $n > 1$ and that the assertion has already been proved for $n-1$. If $\pi_1^n(p),\ldots,\pi_n^n(p)$ are algebraically dependent and $P(x_1,\ldots,x_n)$ is a polynomial of minimal total degree such that $P(\pi_1^n(p),\ldots,\pi_n^n(p)) = 0$, then, by applying the homomorphism ϕ of Lemma 3.21, we have

$$\begin{aligned}\phi(P(\pi_1^n(p),\ldots,\pi_n^n(p)) &= P(\phi(\pi_1^n(p)),\ldots,\phi(\pi_n^n(p)))\\ &= P(\pi_1^{n-1}(p),\ldots,\pi_n^{n-1}(p),0) = 0.\end{aligned}$$

By the inductive hypothesis, we obtain that $P(x_1,\ldots,x_n) = x_n P'(x_1,\ldots,x_n)$, where P' is a polynomial of smaller degree. By (3.27), the element $\pi_n^n(p) = [p]_n$ is not a zero divisor in $\mathcal{H}^n$. Thus, $P'(\pi_1^n(p),\ldots,\pi_n^n(p)) = 0$, a contradiction. □

Spherical Mapping. By Theorem 3.22, every element of a local Hecke–Shimura ring of the general linear group can be written as a polynomial in a finite number of generators. Sometimes one needs to find these polynomials explicitly. However, a direct calculation proves to be rather complicated. To simplify the calculation, it turns out to be convenient to introduce certain mappings of the local Hecke–Shimura rings into rings of symmetric polynomials. Since the polynomials that appear do not necessarily have integral coefficients, it will be reasonable to consider Hecke–Shimura rings over the field $\mathbb{Q}$ of rational numbers in place of the ring $\mathbb{Z}$. We shall denote the rings of double cosets over $\mathbb{Q}$ by the same symbol as before, but with an upper tilde, for example,

$$\widetilde{D}(\Lambda,\Sigma) = D_{\mathbb{Q}}(\Lambda,\Sigma) \quad \text{or} \quad \widetilde{\mathcal{H}}_p = \widetilde{\mathcal{H}}_p^n = D_{\mathbb{Q}}(\Lambda^n,\Sigma_p^n).$$

A matrix $g = (g_{ij}) \in \Sigma^n$ is called (*upper*) *triangular* if $g_{ij} = 0$ for all $i > j$ if, moreover, all diagonal entries g_{ii} are positive, the matrix is called *triangular-plus*. By Lemma 3.18, each left coset $\Lambda^n g \subset \Sigma^n$ contains a triangular-plus representative, which is not unique, in contrast to the reduced representative. On the other hand, a product of triangular-plus matrices is again triangular-plus, which is generally not true for reduced matrices. We fix an order $n \in \mathbb{N}$ and a prime number p. Diagonal entries of a triangular-plus representative contained in a left coset $\Lambda g \subset \Sigma_p = \Sigma_p^n$ are nonnegative powers of p, say, $p^{\delta_1}, p^{\delta_2}, \ldots, p^{\delta_n}$. It is clear that such a diagonal depends only on the left coset. Let $x_1, x_2, \ldots, x_n$ be algebraically independent commuting variables. We set

$$\omega_p^n((\Lambda g)) = \prod_{i=1}^{n} (x_i p^{-i})^{\delta_i}, \tag{3.42}$$

and for a given element $t = \sum_j a_j(\Lambda g_j) \in \widetilde{\mathcal{H}}_p = \widetilde{\mathcal{H}}_p^n$ with $a_j \in \mathbb{Q}$ we define the polynomial

$$\omega_p^n(t) = (\omega_p^n(t))(x_1,\ldots,x_n) = \sum_j a_j \omega_p^n((\Lambda g_j)).$$

Theorem 3.23.

(1) *The mapping*

$$\omega = \omega_p^n : \widetilde{\mathcal{H}}_p^n \mapsto \mathbb{Q}[x_1,\ldots,x_n] \tag{3.43}$$

is a $\mathbb{Q}$*-linear homomorphism of the ring* $\widetilde{\mathcal{H}}_p^n$ *into the ring of polynomials in* $x_1,\ldots,x_n$ *over* $\mathbb{Q}$.

(2) *The* ω*-images of the generators* (3.41) *are given by*

$$\omega(\pi_i^n(p)) = p^{-i(i+1)/2} s_i(x_1,\ldots,x_n) \quad (1 \le i \le n),$$

where

$$s_i(x_1,\ldots,x_n) = \sum_{1 \le \alpha_1 < \cdots < \alpha_i \le n} x_{\alpha_1} \cdots x_{\alpha_i}$$

are the elementary symmetric polynomials.

(3) *The mapping* (3.43) *is an isomorphism of the ring* $\widetilde{\mathcal{H}}_p^n$ *onto the subring of all symmetric polynomials in* $x_1,\ldots,x_n$ *over* $\mathbb{Q}$.

Proof. Since a product of triangular-plus matrices is again triangular-plus, part (1) follows directly from the definitions.

It is not hard to check that the set

$$\{R=(r_{\alpha\beta})\in\mathbb{Z}_n^n | R \text{ is reduced}, \det R = p^i,$$
$$r_{\alpha\alpha}=1 \text{ or } p,\ r_{\alpha\beta}=0 \text{ if } 1\le\alpha<\beta\le n \text{ and } r_{\alpha\alpha}=p\}$$

is the set of reduced representatives in the decomposition of $\pi_i(p)$ into left cosets. Since the number of representatives with $r_{\alpha_1,\alpha_1}=\cdots=r_{\alpha_i,\alpha_i}=p$, where $1\le\alpha_1 <\cdots<\alpha_i\le n$, is clearly $p^{\alpha_1-1}\cdots p^{\alpha_i-i}$, we obtain

$$\omega(\pi_i(p)) = \sum_{1\le\alpha_1<\cdots<\alpha_i\le n} p^{(\alpha_1-1)+(\alpha_2-2)+\cdots+(\alpha_i-i)}(x_{\alpha_1}p^{-\alpha_1})\cdots(x_{\alpha_i}p^{-\alpha_i})$$
$$= p^{-i(i+1)/2}s_i(x_1,\ldots,x_n).$$

Since $\omega([1])=1$, part (3) follows from parts (1) and (2) and well-known properties of symmetric polynomials. □

The theorem reduces calculations in the ring $\widetilde{\mathcal{H}}_p^n$ to calculations with symmetric polynomials; in particular, it allows us to express an element of the ring in terms of the generators (3.41) when its ω-image can be expressed through elementary spherical polynomials. The following exercises illustrate the advantages of spherical representations.

Exercise 3.24.

(1) Show that the formal zeta series of the group GL_n defined by

$$Z^n(s)=\sum_{l=1}^{\infty}\frac{t^n(l)}{l^s}$$

can be expanded into a formal Euler product of the form

$$Z^n(s)=\prod_p Z_p^n(p^{-s}),$$

where p runs through all prime numbers, and where the local zeta series $Z_p^n(v)$ are the formal power series with coefficients $t^n(p^\delta)$,

$$Z_p^n(v)=\sum_{\delta=0}^{\infty}t^n(p^\delta)v^\delta.$$

(2) Prove that for each prime number p, the local zeta series $Z_p^n(v)$ is formally a rational fraction in v; more precisely, the following formal identity holds:

$$Z_p^n(v)=\left(\sum_{i=0}^{n}(-1)^i p^{i(i-1)/2}\pi_i^n(p)\right)^{-1}.$$

[Hint: To prove part (2), show first, using Exercise 3.19, that

$$\sum_{\delta=0}^{\infty}\omega(t_n(p^\delta))v^\delta = \sum_{\delta=0}^{\infty}\left(\sum_{\delta_1+\cdots+\delta_n=\delta} p^{\delta_2+2\delta_3+\cdots+(n-1)\delta_n}(x_1p^{-1})\cdots(x_np^{-n})\right)v^\delta$$
$$= \prod_{i=1}^{n}(1-p^{-1}x_iv) = \left(\sum_{i=1}^{n}(-1)^i p^{-i}s_i(x_1,\ldots,x_n)v^i\right)^{-1}$$
$$= \left(\sum_{i=1}^{n}(-1)^i p^{(i-1)i/2}\omega(\pi_i(p)v^i\right)^{-1}$$

and then use Theorem 3.23 (3).]

Exercise 3.25. Show that every $\mathbb{Q}$-linear ring homomorphism λ of $\widetilde{\mathcal{H}}_p^n$ into $\mathbb{C}$ satisfying $\lambda([p]_n) \neq 0$ has the form

$$t \mapsto \lambda(t) = \omega(t)|_{x_1=\alpha_1,\ldots,x_n=\alpha_n} \qquad (t \in \widetilde{\mathcal{H}}_p^n),$$

where ω is the spherical mapping (3.43) and $\alpha_1,\ldots,\alpha_n$ are nonzero complex numbers determined uniquely up to their order (the parameters of λ).

Exercise 3.26. Let $\lambda : \widetilde{\mathcal{H}}^n \mapsto \mathbb{C}$ be a $\mathbb{Q}$-linear homomorphism such that the restrictions λ_p of λ on $\widetilde{\mathcal{H}}_p$ for all prime p satisfy the assumptions of the previous exercise. Show that the formal zeta series of $\widetilde{\mathcal{H}}^n$ with character λ defined by the Dirichlet series

$$Z^n(s;\lambda) = \sum_{a=1}^{\infty}\frac{\lambda(t(a))}{a^s}$$

has an Euler product expansion of the form

$$Z^n(s;\lambda) = \prod_{p\in\mathbb{P}}\sum_{\delta=0}^{\infty}\frac{\lambda_p(t(p^\delta))}{p^{\delta s}} = \prod_{p\in\mathbb{P}} Z_p^n(p^{-s};\lambda_p),$$

where the local zeta series $Z_p^n(p^{-s};\lambda_p)$ can be written in terms of the parameters $\alpha_1(p),\ldots,\alpha_n(p)$ of λ_p in the form

$$Z_p^n(p^{-s};\lambda_p) = \prod_{i=1}^{n}\left(1-\frac{\alpha_i(p)}{p^{s+1}}\right)^{-1}.$$

3.4 Rings of Double Cosets of the Symplectic Group

We shall present here the basic theory of Hecke–Shimura rings in the symplectic case. Since the case of the full Siegel modular group Γ^n is more transparent and includes main features of the general case, we shall begin with this case, following the pattern of the previous section. We shall be brief, giving hints rather than detailed proofs and leaving the details to the reader as exercises.

Global Rings for the Full Modular Group. Our basic pair in this section is

$$\Gamma = \Gamma^n = \mathrm{Sp}_n(\mathbb{Z}), \quad \Delta = \Delta^n = \left\{ M \in \mathbb{Z}_{2n}^{2n} \;\middle|\; {}^tMJM = \mu(M)J \quad (\mu(M) > 0) \right\},$$

where $J = J_n$, as in Section 1.1, denotes the standard skew-symmetric matrix of order $2n$.

Lemma 3.27. *The pair Γ^n, Δ^n is d-finite for every n.*

Proof. The proof is quite similar to that of Lemma 3.11 if we use the principal congruence subgroups $\Gamma(q) = \Gamma^n(q)$ of Γ defined in Section 1.3 in place of $\Lambda(q)$. □

The lemma allows us to define the Hecke–Shimura ring (3.2) of the pair Γ^n, Δ^n (over $\mathbb{Z}$), which will be denoted by

$$\mathcal{L} = \mathcal{L}^n = D(\Gamma^n, \Delta^n). \tag{3.44}$$

According to Section 3.2, the elements of the form (3.4),

$$T(M) = \tau_\Gamma(M) \qquad (M \in \Delta), \tag{3.45}$$

corresponding bijectively to the double cosets $\Gamma M \Gamma$ contained in Δ form a free basis of the ring $\mathcal{L}$ over $\mathbb{Z}$. The following theorem presents a convenient parametrization of the double cosets in terms of symplectic divisors.

Theorem 3.28. (*The theorem on symplectic divisors.*) *Each of the double cosets $\Gamma M \Gamma$ with $\Gamma = \Gamma^n$ and $M \in \Delta = \Delta^n$ contains the unique representative of the form*

$$\mathrm{sd}(M) = \mathrm{diag}(d_1, \ldots, d_n, e_1, \ldots, e_n), \tag{3.46}$$

where

$$d_i,\, e_j \in \mathbb{N},\; d_ie_i = \mu(M) \quad (i,j = 1,\ldots,n), \quad d_1|d_2|\cdots|d_n|e_n|e_{n-1}|\cdots|e_1. \tag{3.47}$$

We shall prove first an important lemma.

Lemma 3.29. *Let* **m** *be a nonzero integral 2n-column (resp., row). Then there exists a matrix $\gamma \in \Gamma = \Gamma^n$ such that*

$$\gamma \mathbf{m} = {}^t(d,0,\ldots,0) \quad (\textit{resp.}, \mathbf{m}\gamma = (d,0,\ldots,0)),$$

where d is the greatest common divisor of the entries of **m**.

Proof. Note first that by Lemma 1.1, the matrix $J = J_n$ as well as the matrices of the form

$$U(V) = \begin{pmatrix} {}^tV^{-1} & 0 \\ 0 & V \end{pmatrix} \quad \text{and} \quad T(S) = \begin{pmatrix} 1_n & S \\ 0 & 1_n \end{pmatrix}$$

with $V \in \Lambda^n$ and ${}^tS = S \in \mathbb{Z}_n^n$ all belong to Γ. Consider, for example, the case of columns. It follows from Lemma 3.13 that it would be sufficient to show that the set

$\{\gamma\mathbf{m}|\gamma \in \Gamma\}$ contains a column $\mathbf{m}'$ with the zero n-subcolumn $\mathbf{m}'_1$ or $\mathbf{m}'_2$ formed of the first or last n entries, respectively. Assuming the contrary, we conclude that for some of the columns, the minimum $d = d(\mathbf{m}') \geq 1$ of the greatest common divisors of entries of $\mathbf{m}'_1$ and $\mathbf{m}'_2$ is minimal. Then, by Lemma 3.13, we can assume that the corresponding subcolumn, say $\mathbf{m}'_2$, has the form ${}^t(d,0,\ldots,0)$, and on replacing $\mathbf{m}'$ by $\mathbf{m}'' = T(S)\mathbf{m}'$ with a suitable S, we can obviously obtain a column with $d(\mathbf{m}'') < d$, which is a contradiction. □

Proof of Theorem 3.28. In our consideration we shall often divide matrices $M \in \Delta^n$ into standard blocks A, B, C, D of size $n \times n$, so that $M = \left(\begin{smallmatrix} A & B \\ C & D \end{smallmatrix}\right)$, and divide each of the blocks into subblocks with fixed indexing, say $A = \begin{pmatrix} A_1 & A_2 \\ A_3 & A_4 \end{pmatrix}$, whose size is clear from context.

We shall apply induction on n. If $n = 1$, the symplectic case actually coincides with the general linear case for the matrices of second order with positive determinant. Suppose that $n > 1$, and that the claim has already been proved for $n-1$. Let $M' \in \Delta^n$ and let d_1 be the greatest common divisor of entries of M'. Then $M' = d_1 M$, where M is primitive, i.e., has coprime entries, and it suffices to consider only primitive M. Let $\delta = \delta(M)$ denote the minimum of the greatest common divisors of entries of the columns of M. By induction on δ, we shall prove that the double coset $\Gamma M \Gamma$ contains a representative with blocks

$$A = \begin{pmatrix} 1 & 0 \\ 0 & A_4 \end{pmatrix}, \quad B = \begin{pmatrix} 0 & 0 \\ 0 & B_4 \end{pmatrix}, \quad C = \begin{pmatrix} 0 & 0 \\ 0 & C_4 \end{pmatrix}, \quad D = \begin{pmatrix} \mu(M) & 0 \\ 0 & D_4 \end{pmatrix}, \tag{3.48}$$

where $\begin{pmatrix} A_4 & B_4 \\ C_4 & D_4 \end{pmatrix} \in \Delta^{n-1}$. Then application of the inductive hypothesis on n to the last matrix will prove the theorem.

Assume first that $\delta = 1$, and let the index i be minimal among the indices of columns of M with coprime entries. replacing M by MJ, if necessary, we may assume that $i \leq n$. Then on replacing M by $MU({}^tV^{-1})$ with a suitable permutation matrix $V \in \Lambda$, we may assume that $i = 1$. In this case, Lemma 3.29 for the first column of M implies that the left coset ΓM contains a representative whose block A_1 is equal to 1 and the blocks A_3, C_1, and C_3 are zero-blocks. Right multiplication of the representative by $U(W)$ with $W = \begin{pmatrix} 1 & 0 \\ {}^tA_2 & 1_{n-1} \end{pmatrix}$ gives us a matrix whose block A_2 is also a zero-block. Right multiplication of the last matrix by $T(S)$ with $S = \begin{pmatrix} -B_1 & -B_2 \\ -{}^tB_2 & 0 \end{pmatrix}$ leads us to a matrix whose blocks B_1 and B_2 are also zero-blocks. Then it follows from relations (1.1) and (1.2) that the resulting matrix has the form (3.48).

If $\delta > 1$, we can use the same arguments as above to show that the coset $\Gamma M \Gamma$ contains a representative M_0 with blocks $A_1 = \delta$, $A_3 = 0$, $C_1 = 0$, $C_3 = 0$, and such that all of the entries of the blocks A_2, B_1, and B_2 belong to the set $1, 2, \ldots, \delta$. It is clear that $\delta(M_0) \leq \delta$. If $\delta(M_0) = \delta$, then all entries of M_0 are divisible by δ, a contradiction. Hence $\delta(M_0) < \delta$. Repeating the same arguments with respect to M_0 and so on, finally we shall find a representative of the form (3.46).

Uniqueness of the symplectic divisors follows from Theorem 3.12, because the numbers $d_1, \ldots, d_n, e_n, \ldots, e_1$ are elementary divisors of M. □

The matrix $\mathrm{sd}(M)$ of the form (3.46) is called the *matrix of symplectic divisors of* M, and the numbers $d_i = d_i(M)$, $e_j = e_j(M)$ are the *symplectic divisors of* M. It follows that the elements

$$T(d_1,\ldots,d_n,e_1,\ldots,e_n) = T(\mathrm{diag}(d_1,\ldots,d_n,e_1,\ldots,e_n)) \tag{3.49}$$

with positive integers d_i, e_j satisfying (3.47) form a free basis of $\mathcal{L}^n$ over $\mathbb{Z}$. Let us now turn to the multiplicative properties of the ring $\mathcal{L}^n$. To begin with, we have the following theorem.

Theorem 3.30. *The dc-ring* $\mathcal{L}^n = D(\Gamma^n, \Delta^n)$ *is commutative for every* n.

Proof. For further use, we shall give two proofs. First of all, the proof of Theorem 3.14 goes through in the symplectic case too, since by Lemma 1.1, the mapping $M \mapsto {}^tM$ is an antiautomorphism of order two of the pair Γ^n, Δ^n, which does not change diagonal matrices. Second, the mapping

$$M \mapsto M^* = \mu(M)M^{-1} \qquad (M \in \Sigma^n)$$

is clearly also an antiautomorphism of order two of the pair Γ^n, Δ^n, and by Proposition 3.7 and Theorem 3.28, it remains only to note that every element $T = T(D)$ of the form (3.49) satisfies

$$T^* = T(D^*) = T(\mathrm{diag}(e_1,\ldots,e_n,d_1,\ldots,d_n)) = T(J_n D J_n^{-1}) = T, \tag{3.50}$$

since $J_n \in \Gamma^n$. □

As to concrete rules of multiplication in $\mathcal{L}$, we have the following theorem, similar to Theorem 3.15.

Theorem 3.31. *The elements* (3.49) *satisfy the rules*

$$\begin{aligned} T(d_1,\ldots,d_n,e_1,\ldots,e_n)&T(d'_1,\ldots,d'_n,e'_1,\ldots,e'_n) \\ &= T(d_1d'_1,\ldots,d_nd'_n,e_1e'_1,\ldots,e_ne'_n), \end{aligned} \tag{3.51}$$

provided that $\gcd(e_1/d_1, e'_1/d'_1) = 1$. *In particular,*

$$\begin{aligned} \langle d\rangle T(d_1,\ldots,d_n,e_1,\ldots,e_n) &= T(d_1,\ldots,d_n,e_1,\ldots,e_n)\langle d\rangle \\ &= T(dd_1,\ldots,dd_n,de_1,\ldots,de_n), \end{aligned} \tag{3.52}$$

where

$$\langle d\rangle = \langle d\rangle_n = T(d\cdot 1_{2n}) = (\Gamma^n d\cdot 1_{2n}). \tag{3.53}$$

The proof is similar to the proof of Theorem 3.15 and based on the following lemma, similar to Lemma 3.16.

Lemma 3.32. *The elements*

$$T(m) = T^n(m) = \sum_{\substack{d_1e_1=\cdots=d_ne_n=m,\\ d_1|d_2|\cdots|d_n|e_n|e_{n-1}|\cdots|e_1}} T(d_1,\ldots,d_n,e_1,\ldots,e_n) \tag{3.54}$$

with $m \in \mathbb{N}$ have decompositions into left cosets of the form

$$T(m) = \sum_{M\in\Gamma\backslash\Delta,\ \mu(M)=m} (\Gamma M) \tag{3.55}$$

and satisfy the relations

$$T(m)T(m') = T(mm') \quad \text{if } \gcd(m,m') = 1. \tag{3.56}$$

Exercise 3.33. Prove Lemma 3.32 and Theorem 3.31.

Exercise 3.34. Show that the matrices (3.46) of symplectic divisors of every two matrices $M, M' \in \Delta^n$ such that $\gcd(e_1(M)/d_1(M), e_1(M')/d_1(M')) = 1$ satisfy the relation

$$\mathrm{sd}(MM') = \mathrm{sd}(M)\mathrm{sd}(M').$$

Proposition 3.35.

(1) *Each left coset ΓM with $\Gamma = \Gamma^n$ and $M \in \Delta^n$ contains a "triangular" representative of the form*

$$\begin{pmatrix} A & B \\ 0 & D \end{pmatrix}, \tag{3.57}$$

where $0 = 0_n$ is the zero matrix of order n.

(2) *Each left coset ΓM contains a unique representative of the form* (3.57), *where D belongs to a fixed system of representatives of left cosets of Σ^n modulo Λ^n, B lies in a fixed class of the set*

$$B(D) = \left\{ B \in \mathbb{Z}^n_n \;\middle|\; {}^tBD = {}^tDB \right\} \tag{3.58}$$

modulo the equivalence relation

$$B \equiv B' \pmod{D} \Leftrightarrow (B - B')D^{-1} \in \mathbb{Z}^n_n, \tag{3.59}$$

and $A = \mu(M)\,{}^tD^{-1}$.

Proof. In order to prove part (1), we shall apply induction on n. By Lemma 3.29, we may assume that the first column of $M = \left(\begin{smallmatrix} A & B \\ C & D \end{smallmatrix}\right)$ has the form ${}^t(a_{11}, 0, \ldots, 0)$. This shows, in particular, that (1) is true for $n = 1$. Using the relations of Lemma 1.1(4), we conclude that the blocks of M in the notation of the proof of Theorem 3.28 have then the form

$$A = \begin{pmatrix} A_1 & A_2 \\ 0 & A_4 \end{pmatrix}, \quad B = \begin{pmatrix} B_1 & B_2 \\ B_3 & B_4 \end{pmatrix}, \quad C = \begin{pmatrix} 0 & 0 \\ 0 & C_4 \end{pmatrix}, \quad D = \begin{pmatrix} D_1 & 0 \\ D_3 & D_4 \end{pmatrix},$$

where $A_1 = a_{11}$ and the matrix $M' = \begin{pmatrix} A_4 & B_4 \\ C_4 & D_4 \end{pmatrix}$ belongs to Δ^{n-1}. By the inductive hypothesis, there is a matrix $\gamma' = \begin{pmatrix} a' & b' \\ c' & d' \end{pmatrix} \in \Gamma^{n-1}$ such that $c'A_4 + d'C_4 = 0$. Then the matrix $\gamma = \begin{pmatrix} a & b \\ c & d \end{pmatrix}$ with $a = \begin{pmatrix} 1 & 0 \\ 0 & a' \end{pmatrix}$, $b = \begin{pmatrix} 0 & 0 \\ 0 & b' \end{pmatrix}$, $c = \begin{pmatrix} 0 & 0 \\ 0 & c' \end{pmatrix}$, and $d = \begin{pmatrix} 1 & 0 \\ 0 & d' \end{pmatrix}$ belongs to Γ^n, and the matrix γM has "triangular" form.

Part (2) follows easily from the definitions and Lemma 1.1. □

Exercise 3.36. For $D \in \Sigma^n$, show that the number $b(D) = \#(B(D)/\operatorname{mod} D)$ of different residue classes of $B(D)$ modulo the equivalence relation (3.59) is finite and satisfies the rules $b(\lambda D \lambda') = b(D)$ if $\lambda, \lambda' \in \Lambda^n$ and $b(D) = d_1^n d_2^{n-1} \cdots d_n$ if $\operatorname{ed}(D) = \operatorname{diag}(d_1, d_2, \ldots, d_n)$.

Local Rings for the Full Modular Group. Similarly to the case of the general linear group, consideration of the global Hecke–Shimura ring $\mathcal{L}^n$ can be essentially reduced to the study of its *local* subrings.

For a prime number p we define the p-subsemigroup of Δ^n by

$$\Delta_p = \Delta_p^n = \left\{ M \in \Sigma^n \;\middle|\; \mu(M) | p^\infty \right\},$$

where we write $a|b^\infty$ if a divides a power of b. By Lemma 3.27, the pair Γ^n, Δ_p^n is d-finite, and so we can define the Hecke–Shimura ring of the pair over $\mathbb{Z}$,

$$\mathcal{L}_p = \mathcal{L}_p^n = D(\Gamma^n, \Delta_p^n). \tag{3.60}$$

Since $\Delta_p \subset \Delta$, the ring $\mathcal{L}_p$ can be naturally considered as a subring of $\mathcal{L}$, called *local* or *p-local subring of* of $\mathcal{L}$.

Theorem 3.37. *The ring $\mathcal{L}^n$ is generated by the local subrings $\mathcal{L}_p^n$ for all prime numbers p.*

Proof. For a matrix $M \in \Delta^n$ and a prime number p, let us define the *matrix of symplectic p-divisors of M* by

$$\operatorname{sd}_p(M) = \operatorname{diag}(p^{v_p(d_1)}, \ldots, p^{v_p(d_n)}, p^{v_p(e_1)}, \ldots, p^{v_p(e_n)}), \tag{3.61}$$

where $d_i = d_i(M)$ and $e_j = e_j(M)$ are the symplectic divisors (3.46) of M, and where $v_p(r)$ denotes the p-order of a nonzero rational number r. It is clear that $\operatorname{sd}_p(M) \in \Delta_p$, whence $T(\operatorname{sd}_p(M)) \in \mathcal{L}_p$. On the other hand, for a fixed $M \in \Delta$, $\operatorname{sd}_p(M) = 1_{2n}$ for all except a finite number of prime p, and

$$\operatorname{sd}(M) = \prod_{p \in \mathbb{P}} \operatorname{sd}_p(M),$$

whence, by Theorem 3.28 and Theorem 3.31,

$$T(M) = T(\operatorname{sd}(M)) = \prod_{p \in \mathbb{P}} T(\operatorname{sd}_p(M)). \quad \square$$

Let us consider now the structure of the local subrings $\mathcal{L}_p$. Our approach is essentially similar to the case of the general linear group and leads to similar results. This will allow us to make a long story short. The elements

$$T(p^{\delta_1},\dots,p^{\delta_n},p^{\varepsilon_1},\dots,p^{\varepsilon_n}) = T\big(\mathrm{diag}(p^{\delta_1},\dots,p^{\delta_n},p^{\varepsilon_1},\dots,p^{\varepsilon_n})\big), \tag{3.62}$$

where δ_i, ε_j are integers satisfying

$$0 \le \delta_1 \le \dots \le \delta_n \le \varepsilon_n \le \dots \le \varepsilon_1 \text{ and } \delta_1+\varepsilon_1 = \dots = \delta_n+\varepsilon_n, \tag{3.63}$$

span the space $\mathcal{L}_p$ over $\mathbb{Z}$. Such an element is called *primitive* if $\delta_1 = 0$ and *imprimitive* if $\delta_1 \ge 1$. An element $T \in \mathcal{L}_p$ is called *primitive* (resp., *imprimitive*) if it is a linear combination of primitive (resp., imprimitive) elements. It follows from (3.52) that the set $\mathcal{I} = \mathcal{I}_p^n$ of all imprimitive elements of $\mathcal{L}_p$ is the principal ideal generated by $\langle p\rangle = \langle p\rangle_n$, $\mathcal{I} = \langle p\rangle\mathcal{L}_p$. Similarly to Lemma 3.21, we obtain the following result.

Lemma 3.38. *The linear mapping $\psi : \mathcal{L}_p^n \mapsto \mathcal{L}_p^{n-1}$ given on elements of the form (3.62) – (3.63) by*

$$\psi(T(p^{\delta_1},\dots,p^{\delta_n},p^{\varepsilon_1},\dots,p^{\varepsilon_n})) = \begin{cases} T(p^{\delta_2},\dots,p^{\delta_n},p^{\varepsilon_2},\dots,p^{\varepsilon_n}) & \text{if } \delta_1 = 0,\\ 0 & \text{if } \delta_1 > 0,\end{cases}$$

is a ring epimorphism with kernel $\mathcal{I}$.

Exercise 3.39. Prove the lemma.

From this lemma, similarly to the proof of Theorem 3.22, using induction on n, we obtain the following structural theorem.

Theorem 3.40. *For every $n = 1,2,\dots$ and every prime number p, the ring $\mathcal{L}_p^n$ is generated over the subring $\mathbb{Z} \simeq \mathbb{Z}[\langle 1\rangle_n]$ by the element*

$$T(p) = T^n(p) = T(\underbrace{1,\dots,1}_{n},\underbrace{p,\dots,p}_{n}) \tag{3.64}$$

and the elements

$$T_i(p^2) = T_i^n(p^2) = T(\underbrace{1,\dots,1}_{n-i},\underbrace{p,\dots,p}_{i},\underbrace{p^2,\dots,p^2}_{n-i},\underbrace{p,\dots,p}_{i}) \tag{3.65}$$

for $i = 1,\dots,n$; the elements $T(p), T_1(p^2),\dots,T_n(p^2)$ are algebraically independent over $\mathbb{Z}$.

Exercise 3.41. Provide details of the proof of the theorem.

Regular Hecke–Shimura Rings for Congruence Subgroups. We shall see here that the basic structure theorems established above for Hecke–Shimura rings of the full modular group remain essentially true in the case of congruence subgroups.

We recall that a subgroup K of the modular group $\Gamma = \Gamma^n$ is called a congruence subgroup if it contains a principal congruence subgroup $\Gamma(q) = \Gamma^n(q)$ defined above (1.33).

It is easy to check by induction on n that the natural map modulo q,

$$\Gamma^n \mapsto \mathrm{Sp}_n(\mathbb{Z}/q\mathbb{Z}) = \left\{ M \in (\mathbb{Z}/q\mathbb{Z})_{2n}^{2n} \,\middle|\, {}^tMJ_nM \equiv J_n \pmod{q} \right\}, \tag{3.66}$$

is an epimorphism with kernel $\Gamma^n(q)$, whence $\Gamma^n(q)$ is a normal subgroup of the modular group with the factor group isomorphic to $\mathrm{Sp}_n(\mathbb{Z}/q\mathbb{Z})$. In particular, we have the following equality for the index of $\Gamma^n(q)$:

$$\nu(\Gamma^n(q)) = [\Gamma^n : \Gamma^n(q)] = \#(\mathrm{Sp}_n(\mathbb{Z}/q\mathbb{Z})). \tag{3.67}$$

Lemma 3.42. *If q and q' are coprime integers, then $\Gamma^n(q)\Gamma^n(q') = \Gamma^n$.*

Proof. The relation (3.67) shows that the index $\nu(\Gamma^n(q))$ is the number of solutions of a system of polynomial congruences modulo q. Then the Chinese remainder theorem implies that the index is multiplicative in q, i.e., $\nu(\Gamma^n(qq')) = \nu(\Gamma^n(q))\nu(\Gamma^n(q'))$, provided that q and q' are coprime. On the other hand, if q and q' are coprime, then clearly $\Gamma^n(q) \cap \Gamma^n(q') = \Gamma^n(qq')$, which together with the multiplicativity of indexes implies that the embedding $\Gamma^n(q) \subset \Gamma^n$ defines an isomorphism of the factor groups $\Gamma^n(q)/\Gamma^n(qq')$ and $\Gamma^n/\Gamma^n(q')$. Thus, for every $\gamma \in \Gamma^n$ there is $\gamma' \in \Gamma^n(q')$ such that $\gamma\gamma' \in \Gamma(q)$. Then $\gamma \in \Gamma^n(q)(\gamma')^{-1} \subset \Gamma^n(q)\Gamma^n(q')$. □

Let us turn now directly to Hecke–Shimura rings of congruence subgroups. Let $K = K^n$ be a congruence subgroup of $\Gamma = \Gamma^n$. It is easy to see that for every $M \in \Delta = \Delta^n$, the intersection $M^{-1}KM \cap K$ is again a congruence subgroup of Γ. Hence, it has finite index in K. By Lemma 3.3, the double coset KMK is a finite union of left cosets modulo K. Similarly, KMK is a finite union of right cosets modulo K. Thus, the pair K, Δ is d-finite, and we can define the Hecke–Shimura ring $D(K, \Delta)$ of the pair. However, this ring is generally too big to have good properties and is usually replaced by appropriate subrings corresponding to certain subsemigroups of Δ. First of all, for positive integers n and q, we define the subsemigroup

$$\Delta(q) = \Delta^n(q) = \left\{ M \in \Delta \,\middle|\, M \equiv \begin{pmatrix} 1_n & 0 \\ 0 & \mu(M)\cdot 1_n \end{pmatrix} \pmod{q}\ \gcd(\mu(M), q) = 1 \right\} \tag{3.68}$$

of $\Delta = \Delta^n$ and define the dc-ring

$$\mathcal{L}(q) = \mathcal{L}^n(q) = D(\Gamma^n(q), \Delta^n(q)) \tag{3.69}$$

as the *(regular) Hecke–Shimura ring* of the principal congruence subgroup $\Gamma^n(q)$. We shall restrict ourselves to consideration of congruence subgroups K of Γ containing a principal congruence subgroup $\Gamma(q)$ that satisfies the condition

$$K\Delta(q) = \Delta(q)K. \tag{3.70}$$

Such group K will be called a *q-symmetric group of genus n*. For a q-symmetric group of genus n, we set

$$\Delta(K) = \Delta(K,q) = K\Delta(q)K = K\Delta(q) = \Delta(q)K. \tag{3.71}$$

It is clearly a subsemigroup of Δ, and the pair K, $\Delta(K)$ is d-finite. The Hecke–Shimura ring

$$\mathcal{L}(K) = D(K, \Delta(K)) \tag{3.72}$$

will be called the (*regular*) *Hecke–Shimura ring* of the group K. By

$$T_K(M) = \tau_K(M) \qquad (M \in \Delta) \tag{3.73}$$

we shall denote the basic elements of the ring $\mathcal{L}(K)$ of the form (3.4) corresponding to double cosets $KMK \subset \Delta(K)$ and refer to these elements as *double classes*. The following important theorem discloses relations between regular Hecke–Shimura rings for various q-symmetric groups of given genus.

Theorem 3.43. *Let K and K' be two q-symmetric groups of a given genus n. Suppose that $K' \subset K$. Then the following assertions are true:*

(1) $\Delta(K) = K\Delta(K') = \Delta(K')K$.
(2) $\Gamma \cap \Delta(K) = K$.
(3) *If $M, M' \in \Delta(K')$ and $M' \in KMK$, then $M' \in \Gamma(q)MK'$.*
(4) *Let $M \in \Delta(K')$ and let $T_K(M)$ and $T_{K'}(M)$ be two elements of the form* (3.73) *then a decomposition $T_K(M) = \sum_i (KM_i)$ with $M_i \in \Delta(K')$ implies the decomposition $T_{K'}(M) = \sum_i (K'M_i)$ and vice versa.*
(5) *The linear mappings $\iota : \mathcal{L}(K) \mapsto \mathcal{L}(K')$ and $\iota' : \mathcal{L}(K') \mapsto \mathcal{L}(K)$ defined on the double classes by*

$$\iota(T_K(M)) = T_{K'}(M') \quad \text{if } M \in \Delta(K) \text{ and } M' \in KMK \cap \Delta(K')$$

and

$$\iota'(T_{K'}(M')) = T_K(M') \quad \text{if } M' \in \Delta(K'),$$

respectively, are mutually inverse isomorphisms of the rings.

Proof. By (3.71), we get $\Delta(K) = K\Delta(q) = KK'\Delta(q) = K\Delta(K')$. Similarly, $\Delta(K) = \Delta(K')K$. This proves part (1).

If $\gamma \in \Gamma \cap \Delta(K)$, then $\gamma = \delta M$, where $\delta \in K$ and $M \in \Delta(K)$. Hence $M = \delta^{-1}\gamma \in \Gamma \cap \Delta(q) = \Gamma(q)$. Thus, $\gamma \in K\Gamma(q) \subset K$, and part (2) is proved.

Let $M' = \gamma M \gamma'$ with $\gamma, \gamma' \in K$. Let us choose an integer q' prime to q such that the matrix $q'M^{-1}$ is integral. By Lemma 3.42, the matrix γ can be written in the form $\gamma = \gamma_1\gamma_2$, where $\gamma_1 \in \Gamma(q)$ and $\gamma_2 \in \Gamma(q')$. Then we can write $M' = \gamma_1 M \gamma_3$, where $\gamma_3 = M^{-1}\gamma_2 M\gamma'$. Since $\gamma_2 \in \Gamma(q')$, it follows that the matrix γ_3 is integral, and so $\gamma_3 \in \Gamma$. On the other hand, $\gamma_3 = M^{-1}\gamma_1^{-1}M'$. If $M = M_0\delta$ and $M' = M_0'\delta'$, where $M_0, M_0' \in \Delta(q)$ and $\delta, \delta' \in K'$, then $\gamma_3 \equiv \delta^{-1}M_0^{-1}M_0'\delta' \equiv \delta^{-1}\delta' \pmod{q}$. Hence $\gamma_3 \in K'$. This proves part (3).

In order to prove part (4), we note that all of the left cosets $K'M_i$ are distinct and their union contains the double coset $K'MK'$. On the other hand, it follows from part (3) that all of the matrices M_i are contained in the double coset $K'MK'$, which proves the direct assertion of part (4). The inverse assertion follows from the direct one and part (1).

Finally, it follows from parts (1), (2), and (4) that the pairs $K, \Delta(K)$ and $K', \Delta(K')$ satisfy the conditions (3.17) and (3.19). Thus, by Proposition 3.8, the corresponding mapping (3.18) is an isomorphism. It follows from (4) that this mapping coincides with the mapping ι defined in part (5). The mapping ι' is just the inverse of ι. □

The above theorem shows that the Hecke–Shimura rings $\mathcal{L}(K)$ of various q-symmetric groups K are all naturally isomorphic to each other and to the ring $\mathcal{L}(q)$. Thus, any of the groups K can be taken to study the ring. Moreover, the isomorphisms not only establish one-to-one correspondences between the basic elements (3.73) of the rings, but also between the left cosets entering in their decompositions. The last circumstance plays an essential part in the theory of representations of these rings by Hecke operators.

An important example of q-symmetric groups apart from the group $\Gamma(q) = \Gamma^n(q)$ is presented by the group $\Gamma_0(q) = \Gamma_0^n(q)$ defined by (2.2).

Lemma 3.44. *Let*

$$\begin{aligned}\Delta_0(q) &= \Delta_0^n(q)\\ &= \left\{ M = \begin{pmatrix} A & B \\ C & D \end{pmatrix} \in \Delta \;\middle|\; \gcd(\det M, q) = 1, C \equiv 0 \pmod{q} \right\}.\end{aligned} \tag{3.74}$$

Then Proposition 3.35 *remains true for the pair* $\Gamma_0(q), \Delta_0(q)$ *in place of* Γ, Δ, *i.e., each left coset of* $\Delta_0(q)$ *modulo* $\Gamma_0(q)$ *contains a "triangular" representative of the form* (3.57). *Moreover,*

$$\Delta_0(q) = \Gamma_0(q)\Delta(q) = \Delta(q)\Gamma_0(q); \tag{3.75}$$

in particular, the group $\Gamma_0(q)$ *is* q*-symmetric, and* $\Delta(\Gamma_0(q)) = \Delta_0(q)$.

Proof. By Proposition 3.35, there is $\gamma \in \Gamma$ such that the matrix γM has the "triangular" form (3.57). Then obviously, $\gamma \in \Gamma_0(q)$. This proves the first assertion. Let $M = \left(\begin{smallmatrix} A & B \\ C & D \end{smallmatrix}\right) \in \Delta_0(q)$. In order to prove that $M \in \Gamma_0(q)\Delta(q)$, by the first part, we can assume that $C = 0$. Thus by (1.3), we get the relations $A^tD = \mu(M)1_n$ and $A^tB = B^tA$; hence the matrices A and D are invertible modulo q, and the matrix $A^{-1}B$ is symmetric modulo q. It follows that the matrix

$$\gamma = \begin{pmatrix} A^{-1} & -A^{-1}BD^{-1} \\ 0 & \mu D^{-1} \end{pmatrix} = \begin{pmatrix} 1_n & -\mu^{-1}A^{-1}B \\ 0 & 1_n \end{pmatrix} \begin{pmatrix} A^{-1} & 0 \\ 0 & \mu D^{-1} \end{pmatrix},$$

where $\mu = \mu(M)$, is a rational q-integral symplectic matrix with multiplier 1. Since the mapping (3.66) is epimorphic, there exists a matrix $\gamma' \in \Gamma$ such that $\gamma' \equiv \gamma$

(mod q). Then it follows from the definitions that $\gamma' \in \Gamma_0$ and $\gamma' M \in \Delta(q)$. Hence, $\Gamma_0(q) \subset\in \Gamma_0(q)\Delta(q)$. The inverse inclusion is obvious. The second of the decompositions (3.75) follows from the first one by application of the antiautomorphism $M \mapsto \mu(M)M^{-1}$ of the semigroup $\Delta_0(q)$. □

For positive integers n and q, we define the semigroup

$$\Delta_{\langle q\rangle} = \Delta^n_{\langle q\rangle} = \left\{ M \in \Delta^n \;\middle|\; \gcd(q, \det M) = 1 \right\} \tag{3.76}$$

and the Hecke–Shimura ring

$$\mathcal{L}_{\{q\}} = \mathcal{L}^n_{\langle q\rangle} = D(\Gamma^n, \Delta^n_{\langle q\rangle}). \tag{3.77}$$

Since $\Delta_{\langle q\rangle} \subset \Delta$, the ring $\mathcal{L}_{\langle q\rangle}$ can be naturally considered as a subring of the ring $\mathcal{L}$.

Theorem 3.45. *Let K be a q-symmetric subgroup of the modular group $\Gamma = \Gamma^n$. Then the $\mathbb{Z}$-linear mapping*

$$\iota_K : \mathcal{L}(K) \mapsto \mathcal{L}_{\langle q\rangle} \tag{3.78}$$

defined on the elements of the form (3.73) *by*

$$\iota_K(T_K(M)) = T(d_1, \ldots, d_n, e_1, \ldots, e_n) \qquad (M \in \Delta(K)),$$

where $d_i = d_i(M)$ and $e_j = e_j(M)$ are the symplectic divisors of M, is a ring isomorphism. In particular, the ring $\mathcal{L}(K)$ is commutative.

Proof. It follows from Theorem 3.43 and Lemma 3.44 that each of the rings $\mathcal{L}(K)$ is isomorphic to the ring

$$\mathcal{L}_0(q) = \mathcal{L}^n_0(q) = D(\Gamma^n_0(q), \Delta^n_0(q)), \tag{3.79}$$

and in addition, the isomorphism links elements (3.73) of the rings corresponding to matrices from the same double cosets modulo Γ, and hence, with the same symplectic divisors. Therefore, it suffices to prove the theorem only for the group $\Gamma_0(q)$. Note that the group Γ is also q-symmetric and

$$\Delta(\Gamma, q) = \Gamma\Delta(q) = \Delta(q)\Gamma = \Delta_{\langle q\rangle}. \tag{3.80}$$

Thus, Theorem 3.43 can be applied to the groups $K = \Gamma$ and $K' = \Gamma_0(q)$. Since the matrix $\mathrm{sd}(M')$ of symplectic divisors of a matrix $M' \in \Delta_0(q)$ again belongs to $\Delta_0(q)$, it follows from part (3) of this theorem that $\mathrm{sd}(M') \in \Gamma_0(q)M'\Gamma_0(q)$. Hence in part (5) of the theorem we can take $M = M' = \mathrm{sd}(M')$, which completes the proof. □

The above theorem implies that for every q-symmetric subgroup K of Γ, the double class $T_K(M)$ of a matrix $M \in \Delta(K)$ depends only on the symplectic divisors $d_1, \ldots, d_n, e_1, \ldots, e_n$ of M, which can be arbitrary positive integers satisfying the conditions (3.47) and $\gcd(d_i e_i, q) = 1$. This justifies the notation

$$T_K(M) = T_K(sd(M)) = T_K(d_1,\dots,d_n,e_1,\dots,e_n) \quad (M \in \Delta(K)) \tag{3.81}$$

and

$$\begin{aligned} T_K(m) &= \sum_{M \in K\backslash\Delta(K)/K,\, \mu(M)=m} T_K(M) \\ &= \sum_{\substack{d_1e_1=\cdots=d_ne_n=m,\\ d_1|d_2|\cdots|d_n|e_n|e_{n-1}|\cdots|e_1}} T_K(d_1,\dots,d_n,e_1,\dots,e_n), \end{aligned} \tag{3.82}$$

where m is prime to q. From Lemma 3.32, Theorem 3.31, and Theorem 3.45 we obtain the following theorem.

Theorem 3.46. *The elements* (3.81) *of the Hecke–Shimura ring* $\mathcal{L}(K)$ *of a* q*-symmetric group* $K \subset \Gamma = \Gamma^n$ *satisfy the relations*

$$\begin{aligned} T_K(d_1,\dots,d_n,e_1,\dots,e_n)&T_K(d_1',\dots,d_n',e_1',\dots,e_n') \\ &= T_K(d_1d_1',\dots,d_nd_n',e_1e_1',\dots,e_ne_n'), \end{aligned} \tag{3.83}$$

provided that $\gcd(e_1/d_1, e_1'/d_1') = 1$*; in particular,*

$$\begin{aligned} \langle d\rangle_K T_K(d_1,\dots,d_n,e_1,\dots,e_n) &= T_K(d_1,\dots,d_n,e_1,\dots,e_n)\langle d\rangle_K \\ &= T_K(dd_1,\dots,dd_n,de_1,\dots,de_n), \end{aligned} \tag{3.84}$$

where

$$\langle d\rangle_K = T(d\cdot 1_{2n}) = (\Gamma^n d\cdot 1_{2n}). \tag{3.85}$$

The elements (3.82) *satisfy the relations*

$$T_K(m)T_K(m') = T_K(mm') \quad \textit{if } \gcd(m,m') = \gcd(mm',q) = 1. \tag{3.86}$$

Exercise 3.47. Show that the elements (3.82) for $n = 1$ satisfy the relations

$$T_K(m)T_K(m') = \sum_{d|m,m'} d\cdot\langle d\rangle_K T_K(mm'/d^2) \quad \text{if } \gcd(mm',q) = 1,$$

where $\langle d\rangle_K$ are the elements (3.85). Use the relations to show that the formal Dirichlet series with coefficients $T_K(m)$ for m prime to q has a formal Euler factorization of the form

$$\sum_{m\in\mathbb{N},\,\gcd(m,q)=1} \frac{T_K(m)}{m^s} = \prod_{p\in\mathbb{P},\, p\nmid q} \left(T_K(1) - T_K(p)p^{-s} + \langle p\rangle_K p^{1-2s}\right)^{-1}.$$

[Hint: For $K = \Gamma_0^1(q)$, show first that the linear map $\mathcal{L}_0^1(q) \mapsto \mathcal{H}^1 = D(\Lambda^2,\Sigma^2)$ transforming elements $T_K(D)$ to $t_\Lambda(D)$ is an embedding of the rings and then use Exercise 3.19.]

In the same way as was done for the full modular group, consideration of global Hecke–Shimura rings $\mathcal{L}(K)$ for q-symmetric subgroups K of Γ reduces to the study of local subrings. For a prime number p not dividing q, let us introduce the subsemigroup of $\Delta(K)$ given by

$$\Delta_p(K) = \left\{ M \in \Delta(K) \;\middle|\; \mu(M) | p^\infty \right\} \tag{3.87}$$

and the *local subring* $\mathcal{L}_p(K)$ of $\mathcal{L}(K)$ defined by

$$\mathcal{L}_p(K) = D(K, \Delta_p(K)). \tag{3.88}$$

Similarly to the case of full modular group, using corresponding results and Theorem 3.45, we get the following result.

Theorem 3.48. *The ring $\mathcal{L}(K)$ of each q-symmetric group K of genus n is generated by the local subrings $\mathcal{L}_p(K)$ for all prime numbers p not dividing q; each of the local subrings $\mathcal{L}_p(K)$ is naturally isomorphic to the local subring $\mathcal{L}_p = \mathcal{L}_p^n$ of the ring $\mathcal{L} = \mathcal{L}^n$ and generated over $\mathbb{Z} \simeq \mathbb{Z}[\langle 1 \rangle_K]$ by the $n+1$ algebraically independent elements*

$$T_K(p) = T_K(\underbrace{1, \dots, 1}_{n}, \underbrace{p, \dots, p}_{n}) \tag{3.89}$$

and

$$T_{i,K}(p^2) = T_K(\underbrace{1, \dots, 1}_{n-i}, \underbrace{p, \dots, p}_{i}, \underbrace{p^2, \dots, p^2}_{n-i}, \underbrace{p, \dots, p}_{i}) \quad (i = 1, \dots, n). \tag{3.90}$$

The following lemma describes explicit decompositions of certain generators in terms of left cosets, which will be needed below.

Lemma 3.49. *The following decompositions hold for the group $K = \Gamma_0^n(q)$ and each prime number p not dividing q:*

$$T_K(p) = \sum_{i=0}^{n} \sum_{\substack{D \in \Lambda \backslash \Lambda D_i(p) \Lambda \\ B \in B(D)/mod\, D}} \left(K \begin{pmatrix} p\widehat{D} & B \\ 0 & D \end{pmatrix} \right), \tag{3.91}$$

where $\Lambda = \Lambda^n = \mathrm{GL}_n(\mathbb{Z})$, $B(D)$ are the sets (3.58), *residue classes* mod D *are taken modulo the relation* (3.59),

$$D_i(p) = D_i^n(p) = \mathrm{diag}(\underbrace{1, \dots, 1}_{n-i}, \underbrace{p, \dots, p}_{i}), \tag{3.92}$$

and we set

$$\widehat{D} = {}^tD^{-1}; \tag{3.93}$$

in the same notation,

$$T_{n-1,K}(p^2) = \sum_{D\in\Lambda\backslash\Lambda D_{n-1}(p)\Lambda,B}\left(K\begin{pmatrix}p^2\widehat{D} & B\\ 0 & D\end{pmatrix}\right)$$
$$+\sum_{D\in\Lambda\backslash\Lambda pD_1(p)\Lambda,B}\left(K\begin{pmatrix}p^2\widehat{D} & B\\ 0 & D\end{pmatrix}\right)$$
$$+\sum_{D\in\Lambda\backslash\Lambda D_n(p)\Lambda,B}\left(K\begin{pmatrix}p^2\widehat{D} & B\\ 0 & D\end{pmatrix}\right), \tag{3.94}$$

where the matrix B in each of the three sums on the right ranges over all representatives of residue classes $B(D)/\mathrm{mod}\, D$ such that the rank of the matrix $\left(\begin{smallmatrix}p^2\widehat{D} & B\\ 0 & D\end{smallmatrix}\right)$ over the field $\mathbb{F}_p=\mathbb{Z}/p\mathbb{Z}$ of residues modulo p is equal to 1; finally,

$$T_{n,K}(p^2) = \langle p\rangle_K = \left(K(p\cdot 1_{2n})\right). \tag{3.95}$$

Proof. Without loss of generality it suffices to consider the case $K=\Gamma=\Gamma^n$. It follows from Theorem 3.28 that the set $\Delta(p)=\{M\in\Delta^n|\mu(M)=p\}$ consists of the single double coset modulo Γ of the diagonal matrix generating the element (3.89). By (1.2), an integral matrix of the form (3.57) belongs to $\Delta(p)$ if and only if $A=p\widehat{D}$ and $B\in B(D)$. According to Theorem 3.12, the matrix $p\widehat{D}$ with an integral matrix D is integral if and only if D belongs to one of the double cosets $\Lambda D_i(p)\Lambda$ for $i=0,1,\ldots,n$. Thus, the decomposition (3.91) follows from Proposition 3.35(2). Further, it follows from Theorem 3.28 or formula (3.54) that the set $\Delta(p^2)=\{M\in\Delta^n|\mu(M)=p^2\}$ is the union of the double cosets modulo Γ of the matrices generating elements (3.90), and the double coset corresponding to an element $T_i(p^2)$ consists of all matrices in $\Delta(p^2)$ whose rank over the field $\mathbb{F}_p$ is equal to $n-i$. It is easy to see that the rank over $\mathbb{F}_p$ of a matrix in $\Delta(p^2)$ of the form (3.57) can be equal to 1 only if $D\in\Lambda D_i(p)D_j(p)\Lambda$, where $0\le j\le i\le n$ and $j+n-i\le 1$, which implies that $(i,j)=(n,0)$, $(n-1,0)$, or $(n,1)$. Then the decomposition (3.94) follows from Proposition 3.35(2). The decomposition (3.95) is obvious. □

Exercise 3.50. Show that for $K=\Gamma_0^n(q)$ the decompositions of the elements (3.90) with $i=1,\ldots,n$ into left cosets can be written in the form

$$T_i(p^2) = \sum_{\alpha+\beta\le n,\alpha\ge i}\ \sum_{D\in\Lambda\backslash\Lambda D_\alpha(p)D_\beta(p)\Lambda,B}\left(K\begin{pmatrix}p^2\widehat{D} & B\\ 0 & D\end{pmatrix}\right),$$

where the matrix B in each of the inner sums ranges over all representatives of residue classes $B(D)/\mathrm{mod}\, D$ such that the rank of the matrix $\left(\begin{smallmatrix}p^2\widehat{D} & B\\ 0 & D\end{smallmatrix}\right)$ over the field $\mathbb{F}_p$ is equal to $n-i$.

Spherical Mapping. Calculations in local Hecke–Shimura rings of the symplectic group can be considerably simplified if one uses a realization of the rings given by spherical polynomials. Since these polynomials have, generally speaking, not integral but rational coefficients, it will be convenient to extend the rings of double cosets under consideration to the corresponding rings over the field $\mathbb{Q}$ of rational numbers, i.e., the rings

$$\widetilde{\mathcal{L}}(K) = D_{\mathbb{Q}}(K, \Delta(K)), \quad \widetilde{\mathcal{L}}_p(K) = D_{\mathbb{Q}}(K, \Delta_p(K)), \text{ and } \quad \widetilde{\mathcal{L}}_p = \widetilde{\mathcal{L}}_p^n = D_{\mathbb{Q}}(\Gamma^n, \Delta_p^n).$$

It follows from Theorem 3.48 that each of the local rings $\widetilde{\mathcal{L}}_p(K)$, for a q−symmetric group K and a prime p not dividing q, is naturally isomorphic to the ring $\widetilde{\mathcal{L}}_p$. Hence, one can restrict oneself to consideration of spherical mappings of the rings $\widetilde{\mathcal{L}}_p$.

Let $M \in \Delta_p = \Delta_p^n$. According to Lemma 3.35 and (1.3), the left coset ΓM, where $\Gamma = \Gamma^n$, contains a representative of the form

$$M' = \begin{pmatrix} p^\delta \widehat{D} & B \\ 0 & D \end{pmatrix}, \text{ where } p^\delta = \mu(M). \tag{3.96}$$

It is clear that the left coset ΛD modulo $\Lambda = \Lambda^n$ depends only on the left coset ΓM. Let $x_0, x_1, \ldots, x_n$ be independent (commuting) variables. We assign to the left coset (ΓM) the monomial

$$\Omega((\Gamma M)) = x_0^\delta \omega((\Lambda D)),$$

where $\omega((\Lambda D))$ is the monomial in the variables $x_1, \ldots, x_n$ assigned to the left coset (ΛD) by (3.42), and to every element $T = \sum_j a_j(\Gamma M_j) \in \widetilde{\mathcal{L}}_p = \widetilde{\mathcal{L}}_p^n$ with $a_j \in \mathbb{Q}$ we assign the polynomial

$$\Omega(T) = (\Omega_p^n(T))(x_0, x_1, \ldots, x_n) = \sum_j a_j \Omega((\Gamma M_j)) \in \mathbb{Q}[x_0, x_1, \ldots, x_n]. \tag{3.97}$$

We shall call the map $\Omega = \Omega_p^n : \widetilde{\mathcal{L}}_p^n \mapsto \mathbb{Q}[x_0, x_1, \ldots, x_n]$ the *spherical mapping* of the ring $\widetilde{\mathcal{L}}_p^n$.

Theorem 3.51. *The spherical mapping Ω_p^n is a $\mathbb{Q}$-linear isomorphism of the ring $\widetilde{\mathcal{L}}_p^n$ onto the subring $\mathbb{Q}[x_0, x_1, \ldots, x_n]_W$ of $\mathbb{Q}[x_0, x_1, \ldots, x_n]$ consisting of all polynomials invariant under the group $W = W^n$ of transformations of rational fractions in $x_0, x_1, \ldots, x_n$ generated by all permutations in variables $x_1, \ldots, x_n$ and by the transformations $\rho_1, \ldots, \rho_n$ of the variables $x_0, x_1, \ldots, x_n$ defined by*

$$\rho_i(x_0) = x_0 x_i, \quad \rho_i(x_i) = x_i^{-1}, \text{ and } \rho_i(x_j) = x_j \text{ if } j \neq 0, i. \tag{3.98}$$

Proof. In order to prove the theorem, it is sufficient, by Theorem 3.40, to show that the Ω-images of the generators (3.64)–(3.65) are algebraically independent and generate the ring $\mathbb{Q}[x_0, x_1, \ldots, x_n]_W$. Here we shall restrict ourselves to small genera $n = 1$ and $n = 2$, since these are the only cases we use below. The general proof is based on similar ideas but demands more computations and space.

For $n = 1$ it follows from the decompositions of Lemma 3.49 that $\Omega(T(p)) = x_0 + x_0 x_1$ and $\Omega(T_1(p^2)) = \Omega(\langle p \rangle) = p^{-1} x_0^2 x_1$. These polynomials are clearly algebraically independent. Thus, it is sufficient to prove that

$$\mathbb{Q}[x_0,x_1]_W = \mathbb{Q}[x_0 + x_0x_1, x_0^2x_1].$$

The substitution $x_0 = z_1$, $x_0x_1 = z_2$ transforms the ring $\mathbb{Q}[x_0 + x_0x_1, x_0^2x_1]$ into $\mathbb{Q}[z_1 + z_2, z_1z_2]$ and the ring $\mathbb{Q}[x_0,x_1]_W$ into the ring $\mathbb{Q}[z_1,z_2]_S$ of all symmetric polynomials in z_1, z_2 over $\mathbb{Q}$ (note that if a polynomial $P(x_0,x_1) = \sum_{i,j\geq 0} c_{ij}x_0^ix_1^j$ is W-invariant, i.e., $P(x_0x_1, x_1^{-1}) = P(x_0,x_1)$, then $c_{ij} \neq 0$ implies that $i \geq j$, and so the image $P(z_1, z_2/z_1)$ of $P(x_0,x_1)$ is a polynomial in z_1, z_2). Hence, the theorem on symmetric polynomials proves the assertion for $n = 1$.

Standard computations based on the definitions and Proposition 3.35 show that the following formulas hold for $n = 2$:

$$\Omega(T(p)) = x_0(1+x_1)(1+x_2), \tag{3.99}$$

$$\begin{aligned}\Omega(T_1(p^2)) &= x_0^2\omega(t[1,p]) + x_0^2p^3\omega(t[p,p^2]) + x_0^2(p^2-1)\omega(t[p,p])\\ &= \frac{p^2-1}{p^3}x_0^2x_1x_2 + \frac{1}{p}x_0^2(x_1+x_2)(1+x_1x_2),\end{aligned} \tag{3.100}$$

where $t[d,d']$ are the elements (3.25) (note that the number of symmetric matrices of order 2 with rank 1 over the field $\mathbb{F}_p$ is p^2-1), and

$$\Omega(T_2(p^2)) = \Omega(\langle p\rangle) = \frac{1}{p^3}x_0^2x_1x_2. \tag{3.101}$$

According to these formulas, it suffices to show that the polynomials

$$t = x_0(1+x_1)(1+x_2), \quad u = x_0^2x_1x_2, \quad \text{and } v = x_0^2(x_1+x_2)(1+x_1x_2)$$

are algebraically independent and generate the ring $\mathbb{Q}[x_0,x_1,x_2]_W$. Suppose that these polynomials are algebraically dependent, and let $G(y,y_1,y_2)$ be the polynomial of minimal degree such that $G(t,u,v) = 0$. Let us expand G in powers of y_1: $G = \sum_i g_i(y,y_2)y_1^i$. If $g_0 = 0$, then G is divisible by y_1, and $y_1^{-1}G$ is a polynomial of smaller degree with the same property as G. Hence, $g_0 \neq 0$. By the assumption, we have $\sum_i g_i(t,v)u^i = 0$. Since it is an identity in x_0, x_1, x_2, we can set here $x_2 = 0$. The substitution transfers the polynomials t, v into $x_0(1+x_1)$, $x_0^2x_1$, respectively, and u turns to zero. Hence, we get the identity $g_0(x_0(1+x_1), x_0^2x_1) = 0$, a contradiction, because the polynomials $x_0 + x_0x_1$ and $x_0^2x_1$ are algebraically independent. Now we shall prove by induction on m that each $W-$invariant polynomial $F(x_0,x_1,x_2)$ of degree m relative to x_0 is a polynomial in t, u, v. If $m = 0$, then $F = F(x_1,x_2)$ is a symmetric polynomial satisfying $F(x_1^{-1},x_2) = F(x_1,x_2)$, and therefore it must be equal to a constant. Suppose that the assertion has already been proved for the polynomials whose degree in x_0 is less than m with $m \geq 1$. Let

$$F(x_0,x_1,x_2) = \sum_{0\leq i\leq m} x_0^i f_i(x_1,x_2)$$

be $W-$invariant polynomial of degree m relative to x_0. Since the variables x_0, x_1, x_2 are algebraically independent, it follows from the definition of the group W that each of the polynomials $x_0^i f_i(x_1, x_2)$ is also $W-$invariant. Thus, we may assume that $F = x_0^m f(x_1, x_2)$. Since F is W^2-invariant, it is also W^1-invariant, where W^1 acts only on x_0, x_1. In particular, it is true for the polynomial $F(x_0, x_1, 0)$. Since we have already proved the assertion for $n = 1$, we can write $F(x_0, x_1, 0) = P(x_0(1+x_1), x_0^2 x_1)$, where $P(y_1, y_2) = \sum_{i,j} a_{ij} y_1^i y_2^j$ is a polynomial over $\mathbb{Q}$. Since F is of the form $x_0^m f(x_1, x_2)$, and the variables x_0, x_1 are algebraically independent, it follows that $a_{ij} = 0$ if $i + 2j \neq m$. Thus, the polynomial $F'(x_0, x_1, x_2) = F(x_0, x_1, x_2) - P(t, v)$ can be written in the form

$$F'(x_0, x_1, x_2) = x_0^m f(x_1, x_2) - \sum_{i+2j=m} a_{ij} t^i v^j = x_0^m G(x_1, x_2),$$

where G is a polynomial. Since the polynomials t, v turn at $x_2 = 0$ into $x_0(1+x_1)$ and $x_0^2 x_1$, respectively, it follows that the polynomial F' is identically zero at $x_2 = 0$. Thus, it is divisible by x_2. Since F' is symmetric in x_1, x_2, it is divisible by $x_1 x_2$. Hence $F'(x_0, x_1, x_2) = x_0^m x_1 x_2 G'(x_1, x_2)$, where G' is a polynomial. Since F' is W-invariant, it is invariant under the transformation ρ_1, i.e., it satisfies $F'(x_0 x_1, x_1^{-1}, x_2) = F'(x_0, x_1, x_2)$. It follows that G' satisfies $x_1^{m-2} G'(x_1^{-1}, x_2) = G'(x_1, x_2)$. If $m = 0$ or $m = 1$, than G' is not a polynomial in x_1. Thus, $m \geq 2$, and we can write

$$F(x_0, x_1, x_2) = F'(x_0, x_1, x_2) - P(t, v) = x_0^2 x_1 x_2 F''(x_0, x_1, x_2) - P(t, v),$$

where F'' is a $W-$invariant polynomial of degree $m-2$ in x_0. □

The theorem will allow us to reduce calculations in local Hecke–Shimura rings of the symplectic group to calculations with invariant polynomials. In order to illustrate this approach, we shall consider generating formal power series for the sums $T(p^\delta) \in \mathcal{L}_p^n$ of all double cosets of multiplier p^δ. It follows from (3.54) and Proposition 3.35 that

$$T(p^\delta) = \sum_{\substack{0 \leq \delta_1 \leq \cdots \leq \delta_n \leq \delta, \\ D \in \Lambda \backslash \Lambda \operatorname{diag}(p^{\delta_1}, \ldots, p^{\delta_n}) \Lambda, \\ B \in B(D)/\operatorname{mod} D}} \left(\Gamma \begin{pmatrix} p^\delta \widehat{D} & B \\ 0 & D \end{pmatrix} \right), \tag{3.102}$$

where $\Lambda = \Lambda^n$ and $\Gamma = \Gamma^n$. It is easy to check that the number of elements of the set $B(D)/\operatorname{mod} D$ for a matrix $D \in \Sigma^n$ with elementary divisors $d_1, \ldots, d_n$ depends only on the elementary divisors and is equal to $d_1^n d_2^{n-1} \cdots d_n$. Therefore, by the definition of Ω, we obtain the following formula for Ω-images of the elements $T(p^\delta)$:

$$\Omega(T(p^\delta)) = x_0^\delta \sum_{0 \leq \delta_1 \leq \cdots \leq \delta_n \leq \delta} p^{n\delta_1 + (n-1)\delta_2 + \cdots + b_n} \omega(t(p^{\delta_1}, \ldots, p^{\delta_n})), \tag{3.103}$$

where ω is defined by (3.42) and (3.43). Hence, the Ω-image of the generating formal power series of genus n with coefficients $T(p^\delta)$,

$$Z_p(v) = \sum_{\delta=0}^{\infty} T(p^\delta)v^\delta, \tag{3.104}$$

can be written in the form

$$\begin{aligned}\Omega(Z_p)(v) &= \sum_{\delta=0}^{\infty} \Omega(T(p^\delta))v^\delta \\ &= \sum_{0\le\delta_1\le\cdots\le\delta_n\le\delta} p^{n\delta_1+(n-1)\delta_2+\cdots+b_n}\omega(t(p^{\delta_1},\ldots,p^{\delta_n}))(x_0v)^\delta.\end{aligned} \tag{3.105}$$

For the genus $n=1$, we get

$$\begin{aligned}\Omega(Z_p)(v) &= \sum_{0\le\delta_1\le\delta} p^{\delta_1}(x_1p^{-1})^{\delta_1}(x_0v)\delta \\ &= (1-x_0v)^{-1}(1-x_0x_1v)^{-1} = (1-\Omega(T(p))v+p\Omega((p))v^2)^{-1},\end{aligned} \tag{3.106}$$

where $(p)=(p)_1=T_1(p^2)$ is the double coset of the form (3.53) (formulas for $\Omega(T(p))$ and $\Omega(T_1(p^2))$ follow from Proposition 3.35). Hence, since the spherical mapping is monomorphic, we obtain summation formulas for the series $Z_p(v)$.

Proposition 3.52. *For every prime number p, the following identity holds in the ring of formal power series over $\mathcal{L}_p^1$:*

$$\sum_{\delta=0}^{\infty} T(p^\delta)v^\delta = (1-T(p)v+p\langle p\rangle v^2)^{-1}. \tag{3.107}$$

Similarly, for genus $n=2$ we obtain

$$\Omega(Z_p)(v) = \sum_{0\le\delta_1\le\delta_2\le\delta} p^{2\delta_1+\delta_2}\omega(t(p^{\delta_1},p^{\delta_2}))(x_0v)^\delta.$$

Since by (3.27) and Theorem 3.23,

$$\omega(t(p^{\delta_1},p^{\delta_2})) = \omega([p^{\delta_1}]t(1,p^{\delta_2-\delta_1})) = (p^{-3}x_1x_2)^{\delta_1}\omega(t(1,p^{\delta_2-\delta_1})),$$

where $[d]=[d]_2$ are elements (3.28), the sum for $\Omega(Z_p)(v)$ after the replacement of δ_2 and δ by $\delta_1+\alpha$ and $\delta_2+\beta$, respectively, can be rewritten in the form

$$\begin{aligned}\Omega(Z_p)(v) &= \sum_{\alpha,\beta,\delta_1\ge0} p^{2\delta_1+\delta_1+\alpha}(p^{-3}x_1x_2)^{\delta_1}\omega(t(1,p^\alpha))(x_0v)^{\delta_1+\alpha+\beta} \\ &= (1-x_0v)^{-1}(1-x_0x_1x_2v)^{-1}\sum_{\alpha\ge0}\omega(t(1,p^\alpha))(px_0v)^\alpha.\end{aligned}$$

In order to compute the last sum, we note that with the help of Proposition 3.35 it can be easily seen that any set of the form

$$\left\{\begin{pmatrix}1 & a\\ 0 & p^\alpha\end{pmatrix}\ (a \bmod p^\alpha),\ \begin{pmatrix}p^\alpha & 0\\ 0 & 1\end{pmatrix},\ \begin{pmatrix}p^{\alpha-\beta} & b\\ 0 & p^\beta\end{pmatrix}\ (0<\beta<\alpha,\, b \bmod p^\beta,\, p\nmid b)\right\}$$

can be taken as a set of representatives of the left cosets modulo $\Lambda = \Lambda^2$ contained in the double coset $\Lambda \left(\begin{smallmatrix}1 & 0\\ 0 & p^\alpha\end{smallmatrix}\right) \Lambda$. Hence, by the definition of ω, we see that

$$\begin{aligned}
&\sum_{\alpha\geq 0} \omega(t(1,p^\alpha))(px_0v)^\alpha\\
&= \sum_{\alpha\geq 0}\left(p^\alpha(x_2p^{-2})^\alpha + (x_1p^{-1})^\alpha + \sum_{0<\beta<\alpha}(p^\beta - p^{\beta-1})(x_1p^{-1})^{\alpha-\beta}(x_2p^{-2})^\beta\right)(px_0v)^\alpha\\
&= \sum_{\beta,\gamma\geq 0} p^\beta(x_1p^{-1})^\gamma(x_2p^{-2})^\beta(px_0v)^{\beta+\gamma} - p^{-1}\sum_{\beta,\gamma\geq 1} p^\beta(x_1p^{-1})^\gamma(x_2p^{-2})^\beta(px_0v)^{\beta+\gamma}\\
&= (1-x_0x_1v)^{-1}(1-x_0x_2v)^{-1}(1-p^{-1}x_0^2x_1x_2v^2).
\end{aligned}$$

Thus,

$$\begin{aligned}
&\Omega(Z_p)(v)\\
&= [(1-x_0v)(1-x_0x_1v)(1-x_0x_2v)(1-x_0x_1x_2v)]^{-1}(1-p^{-1}x_0^2x_1x_2v^2). \qquad (3.108)
\end{aligned}$$

These calculations allows us to deduce summation formulas for the series $Z_p(v)$ in the case of genus $n = 2$.

Proposition 3.53. *For every prime number p, the following identity holds in the ring of formal power series over* $\mathcal{L}_p^2$*:*

$$\sum_{\delta=0}^{\infty} T(p^\delta)v^\delta = (1-p^2\langle p\rangle v^2)(1-q_1(p)v+q_2(p)v^2-q_3(p)v^3+q_4(p)v^4)^{-1}, \qquad (3.109)$$

where

$$\begin{aligned}
&q_1(p) = T(p), \quad q_2(p) = pT_1(p^2)+p(p^2+1)\langle p\rangle = T(p)^2 - T(p^2) - p^2\langle p\rangle,\\
&q_3(p) = p^3\langle p\rangle T(p), \quad q_4(p) = p^6\langle p\rangle^2.
\end{aligned}$$

Proof. Applying the mapping $\Omega^{-1} : \mathbb{Q}[x_0,x_1,x_2]_W \mapsto \widetilde{\mathcal{L}}_p^2$ coefficient-wise to both side of the identity (3.108), we express the series $Z_p(v)$ in the form of a rational fraction, i.e., a quotient of two polynomials, with numerator $\left(1-\Omega^{-1}(p^{-1}x_0^2x_1x_2)v^2\right)$ and denominator

$$1 - \sum_{1\leq i\leq 4}(-1)^i\Omega^{-1}(s_i(x_0,\, x_0x_1,\, x_0x_2,\, x_0x_1x_2))v^i,$$

where $s_i(y_1,y_2,y_3,y_4)$ are the elementary symmetric polynomials of degree i. By (3.101) we get $\Omega^{-1}(p^{-1}x_0^2x_1x_2) = p^2T_2(p^2) = p^2\langle p\rangle$. By (3.99) we have

$$\begin{aligned}&\Omega^{-1}(s_1(x_0, x_0x_1, x_0x_2, x_0x_1x_2))\\ &\quad = \Omega^{-1}(x_0 + x_0x_1 + x_0x_2 + x_0x_1x_2) = \Omega^{-1}(x_0(1+x_1)(1+x_2)) = T(p).\end{aligned}$$

By (3.101) and (3.100), we deduce

$$\begin{aligned}\Omega^{-1}(s_2(x_0, x_0x_1, x_0x_2, x_0x_1x_2)) &= \Omega^{-1}(x_0^2(x_1 + x_2 + 2x_1x_2 + x_1^2 + x_1x_2^2)\\ &= \Omega^{-1}(2x_0^2x_1x_2) + \Omega^{-1}(x_0^2(x_1+x_2)(1+x_1x_2))\\ &= 2p^3\langle p\rangle + pT_1(p^2) - (p^2-1)p\langle p\rangle\\ &= pT_1(p^2) + (p^2+1)p\langle p\rangle.\end{aligned}$$

The second formula for $q_2(p)$ follows directly by comparing the coefficients of v^2 on both sides of the identity (3.109). By (3.99) and (3.101), we get

$$\begin{aligned}&\Omega^{-1}(s_3(x_0, x_0x_1, x_0x_2, x_0x_1x_2))\\ &\quad = \Omega^{-1}(x_0^2x_1x_2 \cdot x_0(1+x_1+x_2+x_1x_2)) = p^3\langle p\rangle T(p).\end{aligned}$$

Finally, by (3.101),

$$\Omega^{-1}(s_4(x_0, x_0x_1, x_0x_2, x_0x_1x_2)) = \Omega^{-1}(x_0^4x_1^2x_2^2) = p^6\langle p\rangle^2. \quad \square$$

The denominators of the rational functions on the right in (3.106) and (3.108) are special cases of the polynomial

$$\begin{aligned}q(x_0,x_1,\ldots,x_n; v) &= (1-x_0v)\prod_{r=1}^{n}\prod_{1\le i_1<\cdots<i_r\le n}(1 - x_0x_{i_1}\cdots x_{i_r}v)\\ &= \sum_{0\le i\le 2^n}(-1)^i q_i^n(x_0,x_1,\ldots,x_n)v^i.\end{aligned} \tag{3.110}$$

It is easy to see that transformation of the group W^n acting on the variables $x_0,x_1,\ldots,x_n$ only permutes factors of this polynomial. Thus, the coefficients of the expansion of the polynomial in powers of v are invariant under all of the transformations in W^n, and hence, by Theorem 3.51, these coefficients are Ω-images of uniquely determined elements of $\widetilde{\mathcal{L}}_p^n$:

$$q_i^n(x_0,x_1,\ldots,x_n) = \Omega(q_i^n(p)) \quad \text{with } q_i^n(p) \in \widetilde{\mathcal{L}}_p^n. \tag{3.111}$$

The polynomial of degree 2^n over $\widetilde{\mathcal{L}}_p^n$ defined by

$$Q(v) = Q_p^n(v) = \sum_{0\le i\le 2^n}(-1)^i q_i^n(p)v^i \tag{3.112}$$

is called the *spinor p-polynomial of genus n*. It plays an important part in the theory of symplectic Hecke–Shimura rings and Hecke operators. From the obvious relation for polynomials (3.111),

$$q(x_0,x_1,\ldots,x_n;\,v) = v^m (x_0^2 x_1\cdots x_n)^{m/2} q(x_0,x_1,\ldots,x_n;\,(x_0^2x_1\cdots x_n)^{-1}),$$

where $m = 2^n$, there follow the relations for coefficients of the spinor polynomial:

$$q_i^n(x_0,x_1,\ldots,x_n) = (x_0^2x_1\cdots x_n)^{i-m/2} q_{m-i}^n(x_0,x_1,\ldots,x_n) \quad (i = 0,1,\ldots,m = 2^n).$$

Since clearly,

$$\Omega(\langle p\rangle_n) = \Omega(\Gamma(p\cdot 1_{2n})\Gamma) = \Omega(\Gamma(p\cdot 1_{2n})) = p^{-n(n+1)/2} x_0^2 x_1\cdots x_n, \tag{3.113}$$

using (3.111), we see that the coefficients of a spinor p-polynomial of genus n satisfy the following symmetry relations

$$q_i^n(p) = \left(p^{n(n+1)/2}\langle p\rangle_n\right)^{i-m/2} q_{m-i}^n(p) \quad (i = 0,1,\ldots,m = 2^n).$$

In particular,

$$q_{2^n}^n(p) = \left(p^{n(n+1)/2}\langle p\rangle_n\right)^{2^{n-1}}. \tag{3.114}$$

Exercise 3.54. Show that $q_1^n(p) = T^n(p)$.

Spherical mappings enable one to parameterize $\mathbb{Q}$-linear homomorphisms of rings $\widetilde{\mathcal{L}}_p^n$ into the field of complex numbers.

Proposition 3.55. *Each $\mathbb{Q}$-linear homomorphism λ of the ring $\widetilde{\mathcal{L}}_p^n$ into the field $\mathbb{C}$ satisfying $\lambda(\langle p\rangle_n) \neq 0$ has the form*

$$T \mapsto \lambda(T) = \Omega(T)|_{x_0=\alpha_0,x_1=\alpha_1,\ldots,x_n=\alpha_n} \qquad (T \in \mathcal{L}_p^n), \tag{3.115}$$

where Ω is the spherical mapping of $\widetilde{\mathcal{L}}_p^n$ and $\alpha_0,\alpha_1,\ldots,\alpha_n$ are nonzero complex numbers (parameters of λ).

Proof. The spherical mapping of a ring $\widetilde{\mathcal{L}}_p^n$ gives an isomorphism of this ring with the ring of polynomials over $\mathbb{Q}$ generated by the images of the generators (3.64)–(3.65). Hence, it suffices to check that the system of polynomial equations

$$\Omega(T^n(p)) = \beta_0,\, \Omega(T_1^n(p^2)) = \beta_1,\ldots,\Omega(T_1^n(p^2)) = \beta_n$$

has a solution $x_0 = \alpha_0, x_1 = \alpha_1,\ldots,x_n = \alpha_n$ in complex numbers for every $\beta_0,\beta_1,\ldots,\beta_n \in \mathbb{C}$ with $\beta_n \neq 0$. For the same reason as in the proof of Theorem 3.51, we shall restrict ourselves to the simplest cases $n = 1$ and $n = 2$.

If $n = 1$, then computing, with the help of Lemma 3.49, the Ω-images of generators, we obtain the system

$$x_0 + x_0x_1 = \beta_0, \qquad x_0^2x_1 = p\beta_1,$$

which is clearly solvable, since x_0 and x_0x_1 must be equal to the roots of the quadratic equation $X^2 - \beta_0 X + p\beta_1 = 0$. If $n = 2$, then by (3.99)–(3.101), the above system is clearly equivalent to the system

$$x_0(1+x_1+x_2+x_1x_2) = \beta_0, \quad x_0^2(x_1+x_2+x_1^2x_2+x_1x_2^2) = \beta_1', \quad x_0^2x_1x_2 = \beta_2',$$

with $\beta_1' = p\beta_1 - (p^3 - p^2)\beta_2$ and $\beta_2' = p^3\beta_2$. If we set $x_0 + x_0x_1x_2 = X_1$ and $x_0x_1 + x_0x_2 = X_2$, then the first and second equation of the last system turn into $X_1 + X_2 = \beta_0$ and $X_1X_2 = \beta_1'$, respectively. This system of two equations is clearly solvable. Hence we can find the values of the polynomials $x_0 + x_0x_1x_2$ and $x_0x_1 + x_0x_2$. Similarly, the first of these values together with the relation $x_0^2x_1x_2 = \beta_2'$ allows us to determine the values of x_0, $x_0x_1x_2$, and hence the value of x_1x_2. Finally, with these values and the value of $x_0x_1 + x_0x_2$ we can find the values of x_1 and x_2. □

Exercise 3.56. Let $\widetilde{\nu} = \widetilde{\nu}_p^n : \widetilde{\mathcal{L}}_p^n \mapsto \mathbb{Q}$ be the $\mathbb{Q}-$linear extension of the index homomorphism $\nu : \mathcal{L}_p^n \mapsto \mathbb{Z}$ defined in Exercise 3.10. Show that the numbers $1, p, \dots, p^n$ can be taken as parameters of $\widetilde{\nu}$.

3.5 Rings of Triangular-Symplectic Double Cosets

Triangular Rings. It turns out often to be convenient in applications of symplectic Hecke–Shimura rings to split elements of the rings into some elementary components, which, however, do not belong to the initial rings but live in certain extensions. We consider as such extensions Hecke–Shimura rings of the "triangular" subgroup

$$\Gamma_0 = \Gamma_0^n = \left\{ \begin{pmatrix} A & B \\ 0 & D \end{pmatrix} \in \Gamma^n \right\} \tag{3.116}$$

of the modular group. All of the corresponding semigroups will be contained in the semigroup

$$\Delta_0 = \Delta_0^n = \left\{ \begin{pmatrix} A & B \\ 0 & D \end{pmatrix} \in \Delta^n \right\} \tag{3.117}$$

of integral "triangular-symplectic" matrices.

Lemma 3.57. *The pair Γ_0^n, Δ_0^n is d-finite for every n.*

Proof. The proof is similar to that of Lemma 3.11 with the congruence subgroups $\Gamma_0 \cap \Gamma(q)$ in place of $\Lambda(q)$. □

We shall denote by

$$\mathcal{L} = \mathcal{L}^n = D(\Gamma_0^n, \quad \Delta_0^n) \quad \text{and} \quad \mathcal{L}(q) = \mathcal{L}^n(q) = D(\Gamma_0^n, \Delta_0^n \cap \Delta_{\langle q \rangle}^n), \tag{3.118}$$

where $\Delta_{\langle q \rangle}$ are the semigroups (3.76), the Hecke–Shimura ring (3.2) of the group Γ_0^n and the semigroups Δ_0^n and $\Delta_0^n \cap \Delta_{\langle q \rangle}^n$, respectively, and by

$$\mathcal{L}_p = \mathcal{L}_p^n = \mathcal{L}_p^n(q) = D(\Gamma_0^n, \Delta_0^n \cap \Delta_p^n), \tag{3.119}$$

for a prime number p, the local p-subring of $\mathcal{L}$ and $\mathcal{L}^n(q)$ with $p \nmid q$.

Double Cosets and Frobenius Elements. The elements of the form (3.4),

$$\mathbf{T}(M) = \tau_{\Gamma_0}(M) \quad \text{with } M \in \Delta_0, \tag{3.120}$$

corresponding bijectively to the double cosets $\Gamma_0 M \Gamma_0$ contained in Δ_0, form a free basis of $\mathcal{L}$ over $\mathbb{Z}$. From the definition and Proposition 3.35(2) we obtain the following decomposition lemma.

Lemma 3.58. *For each matrix $M = \left(\begin{smallmatrix} A & B \\ 0 & D \end{smallmatrix}\right) \in \Delta_0$, the double coset $\mathbf{T}(M)$ has a decomposition into left cosets of the form*

$$\mathbf{T}(M) = \sum_{\substack{D' \in \Lambda \backslash \Lambda D \Lambda, \\ B' \in B(D', M)/mod\, D'}} \left(\Gamma_0 \begin{pmatrix} \mu \widehat{D}' & B' \\ 0 & D' \end{pmatrix} \right),$$

where $\mu = \mu(M)$, $\widehat{D}' = {}^t(D')^{-1}$, $\Lambda = \Lambda^n$,

$$B(D', M) = \left\{ B' \,\middle|\, \begin{pmatrix} \mu \widehat{D}' & B \\ 0 & D' \end{pmatrix} \in \Gamma_0 M \Gamma_0 \right\},$$

and the equivalence relation modulo D is understood in the sense of (3.59).

In contrast to the symplectic case, the rings $\mathcal{L}$ and $\mathcal{L}_p$ are not commutative and contain zero divisors. For example, one can easily check that for each odd prime p, the elements

$$X = \mathbf{T}\left(\begin{pmatrix} p & 0 \\ 0 & 1 \end{pmatrix} \right), \quad Y = \left(\sum_{i=1}^{p-1} (-1)^i \mathbf{T}\left(\begin{pmatrix} p & i \\ 0 & p \end{pmatrix} \right) \right) \in \mathcal{L}_p^1$$

satisfy the relations $XY = 0$ and $YX \neq 0$. However, certain important properties of symplectic rings remain true also in the triangular-symplectic case. For example, similarly to the relations (3.52) and (3.84), for the same reasons, we have the relations

$$\langle \mathbf{d} \rangle \mathbf{T}(M) = \mathbf{T}(M) \langle \mathbf{d} \rangle = \mathbf{T}(dM) \quad (M \in \Delta_0, d \in \mathbb{N}), \tag{3.121}$$

where

$$\langle \mathbf{d} \rangle = \langle \mathbf{d} \rangle_n = \mathbf{T}(d \cdot 1_{2n}). \tag{3.122}$$

In what follows we shall need only certain subrings and elements of $\mathcal{L}$. The most important of these subrings are the images

$$\begin{aligned} &\mathbf{L} = \iota(\mathcal{L}) \subset \mathcal{L},\ \mathbf{L}_0(q) = \iota(\mathcal{L}_0(q)) \subset \mathcal{L}(q), \\ &\mathbf{L}_p = \iota(\mathcal{L}_p) = \iota(\mathcal{L}_p(\Gamma_0(q)) \subset \mathcal{L}_p, \end{aligned} \tag{3.123}$$

of the symplectic Hecke–Shimura rings $\mathcal{L} = \mathcal{L}^n$, $\mathcal{L}_0(q) = \mathcal{L}_0^n(q)$, $\mathcal{L}_p = \mathcal{L}_p^n$, and $\mathcal{L}_p(\Gamma_0(q)) = \mathcal{L}_p(\Gamma_0^n(q))$ with $p \nmid q$ under embeddings ι of Proposition 3.8 (it follows from Proposition 3.35 that corresponding pairs satisfy conditions (3.17)). We shall refer to these embeddings as *triangular embeddings*. Note that the elements $\iota(T(D))$ with $D \in \Delta_0$, in general, are sums of several double cosets modulo Γ_0 and so differ from single double cosets $\mathbf{T}(D)$. On the other hand, obviously, we have

$$\mathbf{T}(m) = \sum_{M \in \Gamma_0 \backslash \Delta_0, \mu(M)=m} (\Gamma_0 M) = \iota(T(m)) \quad \text{and} \quad \iota(\langle m \rangle) = \langle \mathbf{m} \rangle \quad (m \in \mathbb{N}). \tag{3.124}$$

The mapping $M \mapsto M^* = \mu(M)M^{-1}$ of the semigroup Σ clearly maps the subsemigroup Σ_0 and the subgroup Γ_0 into themselves and so defines an antiautomorphism of order two of the pair Γ_0, Σ_0. Thus, by Proposition 3.8, the mapping

$$\mathbf{T}(M) \mapsto \mathbf{T}^*(M) = \mathbf{T}(M^*) \quad (M \in \Delta_0, M^* = \mu(M)M^{-1}) \tag{3.125}$$

defines an antiautomorphism of the second order of the ring $\boldsymbol{\mathcal{L}}$, which will be referred as the *star mapping*. It is clear that the star mapping transforms each of the local subrings $\boldsymbol{\mathcal{L}}_p$ into itself.

Lemma 3.59. *Each element of the subring $\mathbf{L} \subset \boldsymbol{\mathcal{L}}$ is invariant under the star mapping.*

Proof. If $M \in \Delta_0$, then

$$\iota(T(M))^* = \left(\sum_{M_i \in \Gamma_0 \backslash (\Gamma M \Gamma \cap \Delta_0)/\Gamma_0} \mathbf{T}(M_i) \right)^* = \sum_{M_i \in \Gamma_0 \backslash (\Gamma M \Gamma \cap \Delta_0)/\Gamma_0} \mathbf{T}(M_i^*)$$

$$= \sum_{M_i \in \Gamma_0 \backslash (\Gamma M^* \Gamma \cap \Delta_0)/\Gamma_0} \mathbf{T}(M_i) = \iota(T(M^*)) = \iota(T(M)),$$

since $T(M^*) = T(M)$, by (3.50). □

Along with images of elements of symplectic Hecke–Shimura rings under triangular embeddings, an important role will be played by the *Frobenius elements*

$$\Pi_-(d) = \Pi_-^n(d) = \mathbf{T}\left(\begin{pmatrix} d \cdot 1_n & 0 \\ 0 & 1_n \end{pmatrix} \right)$$

and

$$\Pi_+(d) = \Pi_+^n(d) = \mathbf{T}\left(\begin{pmatrix} 1_n & 0 \\ 0 & d \cdot 1_n \end{pmatrix} \right)$$

parameterized by positive integers a.

Lemma 3.60. *The Frobenius elements have the following decompositions into left cosets modulo Γ_0:*

$$\Pi_-(d) = \left(\Gamma_0 \begin{pmatrix} d\cdot 1_n & 0 \\ 0 & 1_n \end{pmatrix}\right),$$

$$\Pi_+(d) = \sum_{B={}^tB\in\mathbb{Z}_n^n/d\mathbb{Z}_n^n} \left(\Gamma_0 \begin{pmatrix} 1_n & B \\ 0 & d\cdot 1_n \end{pmatrix}\right). \tag{3.126}$$

In addition they satisfy the following rules:

$$\Pi_-(d)\Pi_-(d') = \Pi_-(dd'), \quad \Pi_+(d)\Pi_+(d') = \Pi_+(dd'), \tag{3.127}$$

$$\Pi_-(d)\Pi_+(d) = d^{n(n+1)/2}\langle \mathbf{d}\rangle, \tag{3.128}$$

$$\Pi_-(d)\Pi_+(d') = \Pi_+(d')\Pi_-(d) \quad \text{if } \gcd(d,d') = 1, \tag{3.129}$$

$$\Pi_-(d)^* = \Pi_+(d), \qquad \Pi_+(d)^* = \Pi_-(d), \tag{3.130}$$

where $*$ *is the star mapping.*

Proof. The decompositions (3.126) follow from Lemma 3.58, since in both cases the double coset $\Lambda D\Lambda$ consists of a single left coset, and each of the sets $B(D,M)$ coincides with the set of integral symmetric matrices of order n. The relations (3.127)–(3.130) follow directly from the definitions and decompositions (3.126). For example,

$$\Pi_-(d)\Pi_+(d) = \sum_{B={}^tB\in\mathbb{Z}_n^n/d\mathbb{Z}_n^n} \left(\Gamma_0\, d \begin{pmatrix} 1_n & B \\ 0 & 1_n \end{pmatrix}\right) = d^{n(n+1)/2}\langle \mathbf{d}\rangle. \quad \square$$

The following proposition gives decompositions of the ι-images $\mathbf{T}^n(m)$ of the elements (3.55) or the elements (3.82) for $K = \Gamma_0^n(q)$ with $\gcd(m, q) = 1$ when $n = 1$ and $n = 2$.

Proposition 3.61. *The following relations hold in the triangular rings* $\mathcal{L}^1$ *and* $\mathcal{L}^2$*:*

$$\mathbf{T}^1(m) = \sum_{d,d_1\in\mathbb{N},\, dd_1=m} \Pi_+^1(d)\Pi_-^1(d_1), \tag{3.131}$$

$$\mathbf{T}^2(m) = \sum_{d,d_1,d_2\in\mathbb{N},\, dd_1d_2=m} \Pi_+^2(d)\Pi(d_1)\Pi_-^2(d_2), \tag{3.132}$$

where $\Pi_\pm^n(d)$ *are the Frobenius elements,*

$$\Pi(d) = \Pi_1^2(d) = \mathbf{T}(\mathrm{diag}(d,1,1,d)) = \sum_{\substack{D\in\Lambda\backslash\Lambda\,\mathrm{diag}(1,d)\Lambda,\\ B\in B(D)/\mathrm{mod}\, D}} \left(\Gamma_0 \begin{pmatrix} d\widehat{D} & B \\ 0 & D \end{pmatrix}\right), \tag{3.133}$$

$\Lambda = \Lambda^2$, $B(D)$ *are the sets* (3.58), *and equivalence modulo* D *is understood in the sense of* (3.59).

Proof. It follows from (3.55), Proposition 3.35, and the definitions that

$$\mathbf{T}^1(m) = \sum_{dd_1=m,\, b \bmod d} \left(\Gamma_0 \begin{pmatrix} d_1 & b \\ 0 & d \end{pmatrix}\right)$$

and

$$\begin{aligned}\mathbf{T}^2(m) &= \sum_{dd_1 | m} \sum_{\substack{D\in\Lambda\backslash\Lambda\,\mathrm{diag}(1,d_1)\Lambda,\\ B\in B(dD)/\bmod dD}} \left(\Gamma_0 \begin{pmatrix} (m/d)\widehat{D} & B \\ 0 & dD \end{pmatrix}\right)\\ &= \sum_{dd_1d_2=m} \sum_{\substack{D\in\Lambda\backslash\Lambda\,\mathrm{diag}(1,d_1)\Lambda,\\ B\in B(D)/\bmod dD}} \left(\Gamma_0 \begin{pmatrix} d_1d_2\widehat{D} & B \\ 0 & dD \end{pmatrix}\right).\end{aligned}$$

On the other hand, by (3.126) and the definition of multiplication in rings of double cosets, we have

$$\sum_{d,d_1\in\mathbb{N},\, dd_1=m} \Pi^1_+(d)\Pi^1_-(d_1) = \sum_{dd_1=m} \sum_{b \bmod d} \left(\Gamma_0 \begin{pmatrix} 1 & b \\ 0 & d \end{pmatrix} \begin{pmatrix} d_1 & 0 \\ 0 & 1 \end{pmatrix}\right),$$

which proves the decomposition (3.131). The decomposition (3.133) easily follows from Lemma 3.58. Using this decomposition and decompositions (3.126), we see that the sum on the right in (3.132) is equal to

$$\sum_{dd_1d_2=m} \sum_{\substack{S={}^tS\in\mathbb{Z}^2_2/d\mathbb{Z}^2_2,\\ D\in\Lambda\backslash\Lambda\,\mathrm{diag}(1,d_1)\Lambda,\\ B\in B(D)/\bmod D}} \left(\Gamma_0 \begin{pmatrix} d_1d_2\widehat{D} & B+SD \\ 0 & dD \end{pmatrix}\right).$$

Hence, in order to prove (3.132), it suffices to check that the matrix $B+SD$ ranges over a set of representatives $B(D)/\bmod dD$ when B and S run over the sets $B(D)/\bmod D$ and $S={}^tS\in\mathbb{Z}^2_2/d\mathbb{Z}^2_2$, respectively. We leave this verification to the reader as an easy exercise. □

Lemma 3.62. *The elements* $\Pi(d)$ *satisfy the rules*

$$\Pi(d)^* = \Pi(d) \quad \textit{and} \quad \Pi(d)\Pi(d_1) = \Pi(dd_1) \ \textit{if}\ \gcd(d,d_1)=1, \tag{3.134}$$

*where * stands for the star map.*

Proof. By the definitions we have

$$\Pi(d)^* = \mathbf{T}(\mathrm{diag}(1,d,d,1)) = \mathbf{T}(U\mathrm{diag}(1,d,d,1)U) = \mathbf{T}(\mathrm{diag}(d,1,1,d)) = \Pi(d),$$

where U is the matrix of Γ_0 of the form $\left(\begin{smallmatrix} \varepsilon & 0 \\ 0 & \varepsilon \end{smallmatrix}\right)$ with $\varepsilon = \left(\begin{smallmatrix} 0 & 1 \\ 1 & 0 \end{smallmatrix}\right)$. The rest follows from decompositions (3.133) for $\Pi(d)$ and $\Pi(d_1)$, relations (3.26) for the ring $\mathcal{H}^2$, and easy properties of the sets $B(D)/\bmod D$. □

Exercise 3.63. For $d \in \mathbb{N}$ and $i = 0, 1, \dots, n$, set

$$\Pi_i^n(d) = \mathbf{T}\big(\mathrm{diag}(\underbrace{d,\dots,d}_{n-i},\underbrace{1,\dots,1}_{i},\underbrace{1,\dots,1}_{n-i},\underbrace{d,\dots,d}_{i})\big) \in \boldsymbol{\mathcal{L}}^n,$$

so that $\Pi_0^n(d) = \Pi_-^n(d)$ and $\Pi_n^n(d) = \Pi_+^n(d)$ are the Frobenius elements, and $\Pi_1^2(d) = \Pi(d)$ is the element (3.134). Prove the following properties of elements $\Pi_i^n(d)$:

(1)

$$\Pi_i^n(d) = \sum_{\substack{D \in \Lambda\backslash\Lambda D_i(d)\Lambda,\\ B \in B(D)/\mathrm{mod}\, D}} \left(\Gamma_0 \begin{pmatrix} d\widehat{D} & B \\ 0 & D \end{pmatrix}\right),$$

where $\Lambda = \Lambda^n$, $D_i(d) = D_i^n(d) = \begin{pmatrix} 1_{n-i} & 0 \\ 0 & d \cdot 1_i \end{pmatrix}$, $B(D)$ are the sets (3.58), and $\Gamma_0 = \Gamma_0^n$.

(2) $\Pi_i^n(d)^* = \Pi_{n-i}^n(d)$.

(3) If $d = p$ is a prime number, then $\mathbf{T}^n(p) = \Pi_0^n(p) + \Pi_1^n(p) + \cdots + \Pi_n^n(p)$.

Relations in Local Triangular Rings. In order to derive relations between elements of triangular rings, one has to be able to write these elements in terms of corresponding double or left cosets. Elements of images $\mathbf{L}_p = \mathbf{L}_p^n$ of the local symplectic Hecke–Shimura ring $\mathcal{L}_p = \mathcal{L}_p^n$ or $\mathcal{L}_p(\Gamma_0(q)) = \mathcal{L}_p(\Gamma_0^n(q))$ under the triangular embedding are polynomials in the images

$$\iota(T^n(p)) = \iota(T_K(p)) \quad \text{and} \quad \iota(T_i^n(p^2)) = \iota(T_{i,K}(p^2)) \quad (1 \le i \le n) \tag{3.135}$$

with $K = \Gamma_0^n(q)$ of the generators (3.64)–(3.65) or (3.89)–(3.90) of the rings $\mathcal{L}_p^n$ and $\mathcal{L}_p(\Gamma_0^n(q))$, respectively. The formulas of Propositions 3.52, 3.53, and 3.61 will allow us to express the generators (3.135) for $n = 1$ and $n = 2$ through double cosets.

Lemma 3.64. *The following relations hold in the rings $\mathcal{L}_p^1$ and $\mathcal{L}_p^2$, respectively:*

$$\iota(T^1(p)) = \mathbf{T}^1(p) = \Pi_-^1(p) + \Pi_+^1(p), \; \iota(T_1^1(p^2)) = \langle \mathbf{p} \rangle_1 = p^{-1}\Pi_-^1(p)\Pi_+^1(p), \tag{3.136}$$

and

$$\begin{aligned} &\iota(T^2(p)) = \mathbf{T}^2(p) = \Pi_-^2(p) + \Pi(p) + \Pi_+^2(p), \\ &p\iota(T_1^2(p^2)) = \Pi_-^2(p)\Pi(p) + \Pi(p)\Pi_+^2(p) + \Pi(p)^2 - \Pi(p^2) - p(p+1)\langle \mathbf{p} \rangle_2, \\ &\iota(T_2^2(p^2)) = \langle \mathbf{p} \rangle_2 = p^{-3}\Pi_-^2(p)\Pi_+^2(p). \end{aligned} \tag{3.137}$$

Proof. The relations (3.136) follow from (3.131) and (3.128). The first and the last of the relations (3.137) again follow from (3.131) and (3.128). For computation of the image $\iota(T_1^2(p^2))$, we use the coefficient-wise image under the triangular embedding of the relation (3.109) of Proposition 3.53 in the form

$$\sum_{\delta\geq 0}\mathbf{T}(p^\delta)v^\delta = \left(\sum_{0\leq i\leq 4}(-1)^i\mathbf{q}_i(p)v^i\right)^{-1}(1-p^2\langle\mathbf{p}\rangle v^2),$$

where $\mathbf{T}(p^\delta)=\iota(T^2(p^\delta))=\mathbf{T}^2(p^\delta)$, $\mathbf{q}_i(p)=\iota(q_i^2(p))$, and $\langle\mathbf{p}\rangle=\iota(\langle p\rangle_2)=\langle\mathbf{p}\rangle_2$. Since $\iota(q_1^2(p))=\mathbf{T}(p)$ and $\iota(q_2^2(p))=p\iota(T_1^2(p^2))+p(p^2+1)\langle\mathbf{p}\rangle$, by applying the mapping ι to the second formula for $q_2(p)$ in (3.109), we get the relations

$$\mathbf{T}(p^2)=\mathbf{T}(p)^2-\mathbf{q}_2(p)-p^2\langle\mathbf{p}\rangle=\mathbf{T}(p)^2-p\iota(T_1^2(p^2))-p(p^2+p+1)\langle\mathbf{p}\rangle, \tag{3.138}$$

whence we obtain the formula

$$\iota(T_1^2(p^2))=p^{-1}(\mathbf{T}(p)^2-\mathbf{T}(p^2))-(p^2+p+1)\langle\mathbf{p}\rangle.$$

The formula (3.137) follows then by substitution of the expressions for $\mathbf{T}(p)$ and $\mathbf{T}(p^2)$ given by formulas (3.132). □

This lemma allows us to consider relations of Frobenius elements with elements of symplectic local subrings $\mathbf{L}_p\subset\boldsymbol{\mathcal{L}}_p\subset\boldsymbol{\mathcal{L}}$.

Lemma 3.65. *Each of the Frobenius elements $\Pi_\pm(d)=\Pi_\pm^n(d)$ commutes with each element of the symplectic local subring $\mathbf{L}_p=\mathbf{L}_p^n$ for every prime number p not dividing d and, for $n=2$, satisfies the relations*

$$\Pi_\pm(d)\Pi(\delta)=\Pi(\delta)\Pi_\pm(d)\quad\textit{if}\quad \gcd(d,\delta)=1. \tag{3.139}$$

Proof. It suffices to check that these Frobenius elements commute with the generators (3.135) of the ring $\mathbf{L}_p^n$. Here we shall restrict ourselves to the cases $n=1$ and $n=2$, which will be used below. The general case is similar but requires more preliminary considerations. It follows from formulas (3.136) and (3.137) that all of the generators of the rings $\mathbf{L}_p^1$ and $\mathbf{L}_p^2$ are polynomials in elements $\Pi_\pm^1(p)$ and elements $\Pi_\pm^2(p)$, $\Pi(p)$, $\Pi(p^2)$, respectively. Hence, it follows from (3.129) that the lemma is true for $n=1$, and in order to prove the lemma for $n=2$, it is sufficient to verify (3.139). It follows from (3.133) and (3.126) that these formulas hold for $\Pi_-(d)$, since clearly, $dB(D)/\text{mod}\,D=B(D)/\text{mod}\,D$ if $\gcd(d,\det D)=1$. Applying then the star mapping to both sides and using the relations (3.130), (3.134), we obtain the formulas for $\Pi_+(d)$. □

Spherical Mapping. A useful tool for computations in local triangular rings is given by the natural extension of the spherical mappings of local symplectic rings considered in Section 3.4. Since the formula (3.97) defining the spherical mapping of symplectic rings is written in terms of triangular representatives, we can as well use it for local triangular rings: if

$$\mathbf{T} = \sum_j a_j \left(\Gamma_0 \begin{pmatrix} p^{\delta_j}\widehat{D}_j & B_j \\ 0 & D_j \end{pmatrix} \right) \in \widetilde{\mathcal{L}}_p = \widetilde{\mathcal{L}}_p^n = D_{\mathbb{Q}}(\Gamma_0^n, \Delta_0^n \cap \Delta_p^n) \quad \text{with } a_j \in \mathbb{Q}$$

is an element of a local triangular ring over $\mathbb{Q}$, we define the corresponding spherical polynomial by

$$\Omega(\mathbf{T}) = (\Omega_p^n(\mathbf{T}))(x_0, x_1, \ldots, x_n) = \sum_j a_j x_0^{\delta_j} \omega((\Lambda D_j)) \in \mathbb{Q}[x_0, x_1, \ldots, x_n], \quad (3.140)$$

where, as in formula (3.97), $x_0, x_1, \ldots, x_n$ are independent variables, $\Lambda = \Lambda^n$, and $\omega((\Lambda D))$ are monomials (3.42) in $x_1, \ldots, x_n$. The map $\Omega = \Omega_p^n : \widetilde{\mathcal{L}}_p^n \mapsto \mathbb{Q}[x_0, x_1, \ldots, x_n]$ will be referred to as the *spherical mapping*.

Proposition 3.66. (1) *The spherical mapping $\Omega = \Omega_p^n$ is a $\mathbb{Q}$-linear homomorphism of the ring $\mathcal{L}_p$ into the subring of all polynomials in $n+1$ variables $x_0, x_1, \ldots, x_n$ over $\mathbb{Q}$ symmetric in the variables $x_1, \ldots, x_n$.*

(2) *The restriction of the spherical mapping on the symplectic local subring $\mathbf{L}_p = \mathbf{L}_p^n \subset \mathcal{L}_p$ is an isomorphism of the ring $\mathbf{L}_p$ with the subring $\mathbb{Q}[x_0, x_1, \ldots, x_n]_W$ of invariant polynomials defined in Theorem* 3.51.

Proof. The assertion that spherical mapping is a $\mathbb{Q}$-linear homomorphism into the ring $\mathbb{Q}[x_0, x_1, \ldots, x_n]$ follows directly from the definitions. The assertion that the images are symmetric in the variables $x_1, \ldots, x_n$ needs to be checked only for double cosets $\mathbf{T}(M)$ with $M = \begin{pmatrix} p^{\delta}\widehat{D} & B \\ 0 & D \end{pmatrix} \in \Delta_p$. By Lemma 3.58 and the definition of Ω, we have

$$\begin{aligned} \Omega(\mathbf{T}(M)) &= x_0^{\delta} \#(B(D,M)/\mathrm{mod}\, D) \sum_{D'\Lambda \setminus \Lambda D \Lambda} \omega((\Lambda D')) \\ &= x_0^{\delta} \#(B(D,M)/\mathrm{mod}\, D) \omega(t(D)) \end{aligned} \quad (3.141)$$

(note that the cardinality $\#(B(D,M)/\mathrm{mod}\, D)$ clearly depends only on the double coset $\Lambda D \Lambda$). The last polynomial is symmetric in $x_1, \ldots, x_n$ by Theorem 3.23(3).

Part (2) follows from Theorem 3.51, because the mapping $T \mapsto \mathbf{T} = \iota(T)$ with $T \in \widetilde{\mathcal{L}}_p$ defines an isomorphism of $\widetilde{\mathcal{L}}_p$ on $\mathcal{L}_p$ and $\Omega(\iota(T)) = \Omega(T)$. □

The formula (3.141) allows us to find explicitly the spherical polynomials for a number of elements of triangular rings. In particular, we have the following lemma.

Lemma 3.67. *The following formulas hold:*

$$\Omega(\Pi_-^n(p)) = x_0, \quad \Omega(\Pi_+^n(p)) = x_0 x_1 \cdots x_n, \quad \Omega(\langle \mathbf{p} \rangle_n) = p^{-n(n+1)/2} x_0^2 x_1 \cdots x_n,$$

and

$$\Omega(\Pi(p)) = x_0^2(x_1 + x_2), \quad \Omega(\Pi(p^2)) = x_0^4(x_1^2 + (1 - p^{-1})x_1 x_2 + x_2^2).$$

Proof. All of the formulas except for the last one follow directly from formulas (3.141) and Theorem 3.23(2). The last formula follows, since in this case $\#(B(D,M)/\operatorname{mod} D) = \#(B(D)/\operatorname{mod} D) = p^2$ and one can take

$$\Lambda\backslash\Lambda D\Lambda = \left\{ \begin{pmatrix} p^2 & 0 \\ 0 & 1 \end{pmatrix}, \begin{pmatrix} p & b \\ 0 & p \end{pmatrix} \quad (1 \le b < p), \begin{pmatrix} 1 & b' \\ 0 & p^2 \end{pmatrix} \quad (0 \le b' < p^2) \right\}. \quad \square$$

With the help of the spherical representation and the star map one can derive many new relations in local triangular rings. The next lemma contains the relations that will be needed later.

Lemma 3.68. *The following relations hold in the ring $\mathcal{L}_p^2$:*

$$\Pi_-(p)\Pi(p) = p\mathbf{T}(\operatorname{diag}(p^2,p,1,p)), \quad \Pi(p)\Pi_+(p) = p\mathbf{T}(\operatorname{diag}(p,1,p,p^2)), \tag{3.142}$$

$$\Pi_-(p)\Pi(p)\Pi_+(p) = p^3\langle\mathbf{p}\rangle\Pi(p), \tag{3.143}$$

$$p\iota(T_1^2(p^2)) = \Pi_-(p)\Pi(p) + \Pi(p)\Pi_+(p) + p\Psi(p), \tag{3.144}$$

where

$$\Psi(p) = \sum_{\substack{B={}^tB\in\mathbb{Z}_2^2/p\mathbb{Z}_2^2,\\ r_p(B)=1}} \left(\Gamma_0 \begin{pmatrix} p\cdot 1_2 & B \\ 0 & p\cdot 1_2 \end{pmatrix}\right), \tag{3.145}$$

and $r_p(B)$ is the rank of B over the field $\mathbb{Z}/p\mathbb{Z}$; the elements $\Psi(p)$ satisfy the relations

$$\Psi(p)^* = \Psi(p), \tag{3.146}$$

$$\Pi_-(p)\Psi(p) = (p^2-1)\langle\mathbf{p}\rangle\Pi_-(p), \quad \Psi(p)\Pi_+(p) = (p^2-1)\langle\mathbf{p}\rangle\Pi_+(p). \tag{3.147}$$

Proof. The elements in the right side of relations (3.142), (3.143), and (3.147) are, in fact, integral multiples of single double cosets modulo $\Gamma_0 = \Gamma_0^2$. Thus, in order to prove such a relation, one has to check that each of the left cosets of the product on the left side is contained in the double coset on the right and then to justify the coefficient on the right by comparing the numbers of left cosets on both sides or by computing images of the two sides under the spherical mapping. For example, on the right in (3.143) stands a multiple of the double coset $\langle\mathbf{p}\rangle\Pi(p) = \mathbf{T}(\operatorname{diag}(p^2,p,p,p^2))$, whereas each left coset of the product on the left has the form

$$\begin{pmatrix} p\cdot 1_2 & 0 \\ 0 & 1_2 \end{pmatrix}\begin{pmatrix} p\widehat{D} & B \\ 0 & D \end{pmatrix}\begin{pmatrix} 1_2 & B' \\ 0 & p\cdot 1_2 \end{pmatrix}$$
$$= \begin{pmatrix} 1_2 & pBD^{-1} \\ 0 & 1_2 \end{pmatrix}\begin{pmatrix} p^2\widehat{D} & 0 \\ 0 & D \end{pmatrix}\begin{pmatrix} 1_2 & 0 \\ 0 & p\cdot 1_2 \end{pmatrix}\begin{pmatrix} 1_2 & B' \\ 0 & 1_2 \end{pmatrix},$$

where $D \in \Lambda(\mathrm{diag}(1,p))\Lambda$, $B \in B(D)$, and $B' = {}^tB' \in \mathbb{Z}_2^2$, which clearly belongs to the double coset $\Gamma_0\mathrm{diag}(p^2,p,p,p^2)\Gamma_0$. It follows from the relations of Lemma 3.67 that $\Omega(\Pi_-(p)\Pi(p)\Pi_+(p)) = x_0^4x_1x_2(x_1+x_2)$ and $\Omega(\langle \mathbf{p}\rangle\Pi(p)) = p^{-3}x_0^4x_1x_2(x_1+x_2)$, which justifies the coefficient p^3. The proofs of (3.142) and (3.147) are similar and left to the reader.

It easily follows from formula (3.94) of Lemma 3.49 for $n=2$ and $K=\Gamma^2$ or $\Gamma_0^2(q)$ with $p \nmid q$ and formula (3.145) that

$$\iota(T_1^2(p^2)) = \mathbf{T}(\mathrm{diag}(p^2,p,1,p)) + \mathbf{T}(\mathrm{diag}(p,1,p,p^2)) + \Psi(p),$$

which together with relations (3.142) proves the formula (3.144). This formula implies the relation $\Psi(p) = \iota(T_1^2(p^2)) - p^{-1}(\Pi_-(p)\Pi(p) + \Pi(p)\Pi_+(p))$. It shows that $\Psi(p)$ is a linear combination of double cosets modulo Γ_0, which, according to Lemma 3.59 and formulas (3.130), (1.133), is invariant under the star map. Note that the star map transforms each of the relations (3.142) into the other. The same is true for the relations (3.147). Thus, it suffices to prove one of these pairs of relations. □

Exercise 3.69. Prove the relation

$$p\Psi(p) = \Pi(p)^2 - \Pi(p^2) - p(p+1)\langle \mathbf{p}\rangle.$$

Factorization of Denominators. It follows from Propositions 3.52 and 3.53 that the formal local Z-series of the rings $\mathbf{L}_p^n$ given by

$$\mathbf{Z}_p^n(v) = \sum_{\delta\geq 0} \mathbf{T}^n(p^\delta)v^\delta = \sum_{\delta\geq 0} \iota(T^n(p^\delta))v^\delta, \tag{3.148}$$

for $n=1$ and $n=2$, are formally rational functions,

$$\mathbf{Z}_p^n(v) = \mathbf{P}_p^n(v)\mathbf{Q}_p^n(v)^{-1}, \tag{3.149}$$

where the numerators have the form $\mathbf{P}_p^1(v) = 1$ and $\mathbf{P}_p^2(v) = 1 - p^2\langle \mathbf{p}\rangle_2 v^2$, and the denominators

$$\mathbf{Q}_p^n(v) = \sum_{0\leq i\leq 2^n} (-1)^i \mathbf{q}_i^n(p)v^i = \sum_{0\leq i\leq 2^n} (-1)^i \iota(q_i^n(p))v^i \tag{3.150}$$

are polynomials over $\mathcal{L}_p^n$ of degree 2^n. Here we shall obtain a factorization of the denominators.

Proposition 3.70. *The polynomials $\mathbf{Q}_p^n(v)$ with $n=1$ and $n=2$ have the following factorization over $\mathcal{L}_p^n$:*

$$\mathbf{Q}_p^1(v) = (1-\Pi_-^1(p)v)(1-\Pi_+^1(p)v), \tag{3.151}$$

$$\mathbf{Q}_p^2(v) = (1-\Pi_-^2(p)v)(1-\Pi(p)v + p(\langle \mathbf{p}\rangle_2 + \Psi(p))v^2)(1-\Pi_+^2(p)v), \tag{3.152}$$

where $\Psi(p)$ is the element (3.145).

Proof. By formulas (3.107) and (3.136), we have

$$\mathbf{Q}_p^1(v) = 1-(\Pi_-^1(p)+\Pi_+^1(p))v+(\Pi_-^1(p)\Pi_+^1(p))v^2 = (1-\Pi_-^1(p)v)(1-\Pi_+^1(p)v).$$

This proves (3.151). It follows from formulas (3.109) and (3.132) with $a=p$ that

$$\begin{aligned}\mathbf{Q}_p^2(v) =& 1-(\Pi_-+\Pi+\Pi_+)v+(p\iota(T_1)+p(p^2+1)\langle\mathbf{p}\rangle)v^2\\ &-p^3\langle\mathbf{p}\rangle(\Pi_-+\Pi+\Pi_+)v^3+p^6\langle\mathbf{p}\rangle^2v^4,\end{aligned}$$

where $\Pi_\pm=\Pi_\pm^2(p)$, $\Pi=\Pi(p)$, $T_1=T_1^2(p^2)$, and $\langle\mathbf{p}\rangle=\langle\mathbf{p}\rangle_2$. Hence, in order to prove the factorization (3.152) it suffices to verify the relations

$$p\iota(T_1)+p(p^2+1)\langle\mathbf{p}\rangle = \Pi_-\Pi+\Pi_-\Pi_++\Pi\Pi_++p(\langle\mathbf{p}\rangle+\Psi), \tag{3.153}$$

$$p^3\langle\mathbf{p}\rangle(\Pi_-+\Pi+\Pi_+) = p\Pi_-(\langle\mathbf{p}\rangle+\Psi)+p(\langle\mathbf{p}\rangle+\Psi)\Pi_++\Pi_-\Pi\Pi_+, \tag{3.154}$$

and

$$p^6\langle\mathbf{p}\rangle^2 = p\Pi_-(\langle\mathbf{p}\rangle+\Psi)\Pi_+, \tag{3.155}$$

where $\Psi=\Psi(p)$. By formula (3.144), we get

$$p\iota(T_1)+p(p^2+1)\langle\mathbf{p}\rangle = \Pi_-\Pi+\Pi\Pi_++p\Psi+p(p^2+1)\langle\mathbf{p}\rangle,$$

which proves the formula (3.153), since $\Pi_-\Pi_+=p^3\langle\mathbf{p}\rangle$. By formulas (3.147) and (3.144) we obtain that the sum on the right in (3.154) is equal to

$$p(\langle\mathbf{p}\rangle\Pi_-+(p^2-1)\langle\mathbf{p}\rangle\Pi_-)+p(\langle\mathbf{p}\rangle\Pi_++(p^2-1)\langle\mathbf{p}\rangle\Pi_+)+p^3\langle\mathbf{p}\rangle\Pi,$$

which proves formula (3.154). Finally, by (3.147) we have $\Pi_-\Psi\Pi_+=p^3(p^2-1)\langle\mathbf{p}\rangle^2$, which implies formula (3.155). □

The following lemma will be useful in Chapter 5.

Lemma 3.71. *The generating power series for the elements* $\Pi(p^\delta)$ *with a prime* p *satisfy the formal identity*

$$\sum_{\delta\ge0}\Pi(p^\delta)v^\delta = (1-p^2\langle\mathbf{p}\rangle_2v^2)\big(1-\Pi(p)v+p(\langle\mathbf{p}\rangle_2+\Psi(p))v^2\big)^{-1}. \tag{3.156}$$

Proof. By (3.148), (3.149), and (3.152), we have the formal identity

$$\begin{aligned}\sum_{\delta\le0}\mathbf{T}^2(p^\delta)v^\delta &= (1-p^2\langle\mathbf{p}\rangle_2v^2)\\ &\quad\times\big((1-\Pi_-^2(p)v)(1-\Pi(p)v+p(\langle\mathbf{p}\rangle_2+\Psi(p))v^2)(1-\Pi_+^2(p)v)\big)^{-1}\\ &= (1-p^2\langle\mathbf{p}\rangle_2v^2)\left(\sum_{\delta\ge0}\Pi_+(p^\delta)v^\delta\right)\big(1-\Pi(p)v+p(\langle\mathbf{p}\rangle_2+\Psi(p))v^2\big)^{-1}\\ &\quad\times\left(\sum_{\delta\le0}\Pi_-(p^\delta)v^\delta\right).\end{aligned}$$

On the other hand, by formulas (3.132) for $m = p^\delta$, we obtain

$$\sum_{\delta\le 0} \mathbf{T}^2(p^\delta)v^\delta = \sum_{\delta_1,\delta_2,\delta_3\ge 0} \Pi_+(p^{\delta_1})\Pi(p^{\delta_2})\Pi_-(p^{\delta_3})v^{\delta_1+\delta_2+\delta_3}$$
$$= \left(\sum_{\delta\ge 0}\Pi_+(p^\delta)v^\delta\right)\left(\sum_{\delta\ge 0}\Pi(p^\delta)v^\delta\right)\left(\sum_{\delta\ge 0}\Pi_-(p^\delta)v^\delta\right).$$

Since the series $\sum_{\delta\ge 0}\Pi_\pm(p^\delta)v^\delta$ are invertible over the ring of formal power series, the formula (3.156) follows by comparison of the above formulas. □

Note that the summation formula (3.156) is clearly equivalent to the recurrence relations

$$\Pi(p^{\delta+1}) - \Pi(p^\delta)\Pi(p) + p\Pi(p^{\delta-1})(\langle\mathbf{p}\rangle_2 + \Psi(p)) = \begin{cases} -p^2\langle\mathbf{p}\rangle_2 & \text{if } \delta = 1,\\ 0 & \text{if } \delta \ge 1. \end{cases} \tag{3.157}$$

Exercise 3.72. Prove the following formal identities for formal Dirichlet series and Euler products over the rings $\mathcal{L}^1$ and $\mathcal{L}^2$:

$$\sum_{m\in\mathbb{N}} \frac{\mathbf{T}^1(m)}{m^s} = \prod_{p\in\mathbb{P}} \left\{\left(1-\frac{\Pi^1_-(p)}{p^s}\right)\left(1-\frac{\Pi^1_+(p)}{p^s}\right)\right\}^{-1},$$

$$\sum_{m\in\mathbb{N}} \frac{\mathbf{T}^2(m)}{m^s} = \prod_{p\in\mathbb{P}} \left(1-\frac{p^2\langle\mathbf{p}\rangle_2}{p^{2s}}\right)$$
$$\times\left\{\left(1-\frac{\Pi^2_-(p)}{p^s}\right)\left(1-\frac{\Pi(p)}{p^s}+\frac{p(\langle\mathbf{p}\rangle_2+\Psi(p))}{p^{2s}}\right)\left(1-\frac{\Pi^2_+(p)}{p^s}\right)\right\}^{-1}.$$

Chapter 4
Hecke Operators

4.1 Hecke Operators for Congruence Subgroups

As has been indicated in Section 3.1, the main reason for the introduction of Hecke operators was to reveal and explain multiplicative properties of Fourier coefficients of modular forms. The general scheme looks as follows: Hecke–Shimura rings of symplectic groups operate on spaces of modular forms by Hecke operators, and the multiplicative properties of Fourier coefficients of modular forms appear just as reflections of the multiplicative relations in the corresponding Hecke–Shimura rings. In this section we shall first define Hecke operators on an abstract level and then consider Hecke operators for congruence subgroups of the modular group including the questions of invariant subspaces and diagonalization of regular operators.

Abstract Hecke Operators. The definition of Hecke operators on an abstract level looks quite simple and natural. Let $D(\Lambda, \Sigma)$ be a Hecke–Shimura ring (3.2), and $\mathbf{V}$ a $\mathbb{Z}$-module, where the semigroup Σ acts by linear operators $\Sigma \ni g : \mathbf{v} \mapsto \mathbf{v}|g$ satisfying $\mathbf{v}|g|g' = \mathbf{v}|gg'$ for all $\mathbf{v} \in \mathbf{V}$ and $g, g' \in \Sigma$. Let

$$\mathbf{V}(\Lambda) = \left\{ \mathbf{v} \in \mathbf{V} \,\middle|\, \mathbf{v}|\lambda = \mathbf{v}, \quad \forall \lambda \in \Lambda \right\} \tag{4.1}$$

be the submodule of all Λ-invariant elements of $\mathbf{V}$.

Lemma 4.1. *For every $\mathbf{v} \in \mathbf{V}(\Lambda)$ and $\tau = \sum_i a_i(\Lambda g_i) \in D(\Lambda, \Sigma)$, the element*

$$\mathbf{v}|\tau = \sum_i a_i \mathbf{v}|g_i \tag{4.2}$$

does not depend on the choice of representatives $g_i \in \Lambda g_i$; it again belongs to $\mathbf{V}(\Lambda)$, and the operators $|\tau$ satisfy the rules

$$|a\tau + b\tau' = a|\tau + b|\tau', \quad |\tau\tau' = |\tau|\tau' \quad (a, b \in \mathbb{Z},\, \tau, \tau' \in D(\Lambda, \Sigma)). \tag{4.3}$$

A. Andrianov, *Introduction to Siegel Modular Forms and Dirichlet Series*, Universitext,
DOI 10.1007/978-0-387-78753-4_4,

Proof. If $g_i' = \lambda_i g_i$ with $\lambda_i \in \Lambda$ are other representatives, then

$$\sum_i a_i \mathbf{v}|\lambda_i g_i = \sum_i a_i \mathbf{v}|\lambda_i|g_i = \sum_i a_i \mathbf{v}|g_i = \mathbf{v}\tau,$$

by definition of $\mathbf{V}(\Lambda)$. If $\lambda \in \Lambda$, then

$$(\mathbf{v}|\tau)|\lambda = \sum_i a_i \mathbf{v}|g_i\lambda = \mathbf{v}|\tau\lambda = \mathbf{v}|\tau,$$

by the definition of $D(\Lambda, \Sigma)$. The rules (4.3) easily follow from the definitions. □

Thus, the mapping $\tau \mapsto |\tau$ is a linear representation of the ring $D(\Lambda, \Sigma)$ on the module $\mathbf{V}(\Lambda)$. The operators $|\tau$ are called the *Hecke operators*.

Hecke Operators for Congruence Subgroups. Here we shall define Hecke operators on spaces $\mathfrak{M}_k(K, \chi)$ of modular forms of integral weight k and character χ for a congruence subgroup K of the modular group $\Gamma = \Gamma^n$ (see Section 1.3). Let $\Delta' \subset \Delta = \Delta^n$ be a subsemigroup of the semigroup of integral symplectic matrices of order n with positive multipliers that satisfies the condition $K\Delta'K = \Delta'$ and contains the identity matrix 1_{2n}. Suppose that χ is a congruence character of K extendable to a multiplicative homomorphism $\boldsymbol{\chi}$ of the semigroup Δ' into the nonzero complex numbers. Then we can define the action of the semigroup Δ' on the space $\mathfrak{M} = \mathfrak{M}^n$ of holomorphic functions on the upper half-plane $\mathbb{H} = \mathbb{H}^n$ by the linear operators

$$\Delta' \ni M : F \mapsto F|M = F|_{k,\chi}M = \overline{\boldsymbol{\chi}}(M)F|_k M \qquad (F \in \mathfrak{M}), \tag{4.4}$$

generalizing the Petersson operators (1.29), where $\overline{\boldsymbol{\chi}}$ is the complex conjugate character. These operators transfer the space $\mathfrak{M}$ into itself, and by (1.31), satisfy

$$F|MM' = \overline{\boldsymbol{\chi}}(MM')F|_k MM' = \overline{\boldsymbol{\chi}}(M)\overline{\boldsymbol{\chi}}(M')(F|_k M)|_k M' = (F|M)|M'.$$

Thus, the operators define a linear representation of Δ' on $\mathfrak{M}$. By Lemma 4.1, we can define the linear representation of the Hecke–Shimura ring $\mathcal{L}' = D(K, \Delta')$ on the subspace $\mathfrak{M}(K)$ of the form (4.1) of K-invariant holomorphic functions given by Hecke operators (4.2) corresponding to elements $T = \sum_i a_i(KM_i) \in \mathcal{L}'$,

$$F|T = F|_{k,\chi}T = \sum_i a_i F|_{k,\chi}M_i. \tag{4.5}$$

It is easy to see that the definition is independent of the choice of extension $\boldsymbol{\chi}$.

Lemma 4.2. *Under the above assumptions, the subspaces $\mathfrak{M}_k(K, \chi)$ and $\mathfrak{N}_k(K, \chi)$ of modular forms and cusp forms are invariant under all of the Hecke operators (4.5).*

Proof. The condition $F \in \mathfrak{M}(K)$ for a function F holomorphic on $\mathbb{H}$ means that

$$F|_{k,\chi}M = \overline{\chi}(M)F|_k M = F \text{ or } F|_k M = \chi(M)F, \text{ for all } M \in K, \tag{4.6}$$

since the values of χ on K form a finite group of nonzero complex numbers and so belong to the unit circle $\{z \in \mathbb{C} | z\bar{z} = 1\}$. Hence, for genera $n > 1$, we conclude that the subspace $\mathfrak{M}_k(K,\chi)$ coincides with the whole of $\mathfrak{M}(K)$ and so is invariant under all of the Hecke operators. For $n=1$, we have still to check that each of the functions $F|T$ satisfies condition (iii) of the definition of modular forms, if F does. This follows from Proposition 1.21, since $F|T$ is a linear combination of functions $F|_k M_i$ with $M_i \in \Delta$. Invariance of the subspace of cusp forms follows from Proposition 1.25. □

Invariant Subspaces. Here we introduce certain spaces of modular forms invariant under all regular Hecke operators. For positive integers n and q, let K be a q-symmetric group of genus n, i.e. a subgroup of Γ^n containing $\Gamma^n(q)$ and satisfying the condition (3.70). We shall say that a character χ of K that is trivial on $\Gamma^n(q)$ is *extendable* if it can be extended to a homomorphism into the unit circle of the semigroup $\Delta(K)$ defined by (3.71). Below we shall call this homomorphism just an *extension of* χ and denote it by the same symbol χ. Here we shall construct for such K and χ subspaces of the spaces $\mathfrak{M}_k(K,\chi)$ invariant under all of the *regular Hecke operators*, i.e., Hecke operators corresponding to elements of the regular Hecke–Shimura ring $\mathcal{L}(K)=D(K,\Delta(K))$.

In order to introduce invariant subspaces, we define a multiplicative family of linear operators commuting with Hecke operators. First, for an integer d prime to q we denote by $u(d)$ a matrix of Γ^n satisfying the congruence

$$u(d)=u_n(d)\equiv\begin{pmatrix} d^{-1}\cdot 1_n & 0\\ 0 & d\cdot 1_n\end{pmatrix} \pmod q, \tag{4.7}$$

which exists because the mapping (3.66) is epimorphic. It is clear that all matrices $u(d)$ with d belonging to a fixed residue class modulo q belong to a fixed left (= right, = double) coset of Γ modulo $\Gamma(q)$.

Lemma 4.3. *Let K be a q-symmetric group of genus n, and d an integer prime to q. Then the mapping*

$$\Delta(K)\ni M\mapsto M'=u(d)^{-1}Mu(d), \tag{4.8}$$

where $\Delta(K)$ is the semigroup (3.71), *defines an automorphism of $\Delta(K)$ transforming both the subgroup K and each of the double cosets $KMK\subset\Delta(K)$ into themselves; if, in addition, χ is an extendable character of K, then the map* (4.8) *preserves any extension χ of the character to the semigroup $\Delta(K)$, that is, $\chi(M')=\chi(M)$ for all $M\in\Delta(K)$.*

Proof. The map (4.8) is certainly a homomorphism of $\Delta(K)$ onto the semigroup $u(d)^{-1}\Delta(K)u(d)$. Since clearly the matrix $N=du(d)$ belongs to the semigroup $\Delta(q)$ defined by (3.68) and $\Gamma(q)N\Gamma(q)=\Gamma(q)N=N\Gamma(q)$, we conclude, by Theorem 3.43(4) with $K'=\Gamma(q)$, that $KNK=KN=NK$. Hence, for each $\gamma\in K$, there is a unique $\gamma'\in K$ such that $\gamma N=N\gamma'$ and conversely, moreover, $\gamma'=N^{-1}\gamma N=u(d)^{-1}\gamma u(d)$. Hence, we have $u(d)^{-1}Ku(d)=K$. From the definitions, we get $u(d)^{-1}\Delta(q)u(d)=\Delta(q)$. Thus

$$u(d)^{-1}\Delta(K)u(d)=u(d)^{-1}K\Delta(q)u(d)=Ku(d)^{-1}\Delta(q)u(d)=K\Delta(q)=\Delta(K).$$

The matrices M and M' of $\Delta(K)$ belong to the same double coset modulo Γ and so have equal symplectic divisors. Then, by Theorem 3.45, these matrices are in one double coset modulo K. Finally, for $N = du(d)$ we have

$$\chi(M') = \chi(N^{-1}MN) = \chi(N)^{-1}\chi(M)\chi(N) = \chi(M),$$

because $N \in \Delta(K)$ and so $\chi(M^{-1}) = \chi(M)^{-1}$. □

It follows from properties of Petersson operators that the operators $|_k u(d)$ on the space $\mathfrak{M}_k(\Gamma(q), 1)$ with an integral k map the space into itself, depend only on d modulo q, and, since $u(d)u(d') \equiv u(dd') \pmod{q}$, satisfy the rule

$$|_k u(d)|_k u(d') = |_k u(d)u(d') = |_k u(dd'). \tag{4.9}$$

Proposition 4.4. *Let K be a q-symmetric group of genus n, χ an expandable character of K, and k an integer. Then for each integral d coprime to q, the following assertions hold:*

(1) *The subspaces $\mathfrak{M}_k(K, \chi)$ and $\mathfrak{N}_k(K, \chi)$ of the space $\mathfrak{M}_k(\Gamma(q), 1)$ are invariant with respect to the operator $|_k u(d)$.*
(2) *The operator $|_k u(d)$ on $\mathfrak{M}_k(K, \chi)$ commutes with each of the regular Hecke operators $|_{k,\chi} T$.*

Proof. By Proposition 1.25, it suffices to prove the part (1) only for the space $\mathfrak{M}_k(K, \chi)$. In this case, the assertion means that the operator $|_k u(d)$ transforms functions F of $\mathfrak{M}_k(\Gamma(q), 1)$ satisfying the rules $F|_k\gamma = \chi(\gamma)F$ for all $\gamma \in K$ to functions again satisfying these rules, but this follows from Lemma 4.3 and the property (1.31) of Petersson operators.

It suffices to prove part (2) for operators corresponding to elements of the regular ring $\mathcal{L}(K)$ of the form $T_K(M)$ with $M \in \Delta(K)$ defined by (3.73). Since by Lemma 4.3, the mapping (4.8) carries both the group K and each double coset KMK into themselves, it transfers each set $M_1, \ldots, M_\nu$ of representatives of left cosets modulo K contained in KMK to another such set of representatives $M'_1, \ldots, M'_\nu$. Hence by (4.5) and (4.4), for $F \in \mathfrak{M}_k(K, \chi)$ we obtain

$$\begin{aligned} F|_k u(d)|_{k,\chi} T_K(M) &= \sum_{1\le i\le \nu} \overline{\chi}(M'_i)F|_k u(d)|_k M'_i = \sum_{1\le i\le \nu} \overline{\chi}(M_i)F|_k(u(d)M'_i) \\ &= \sum_{1\le i\le \nu} \overline{\chi}(M_i)F|_k M_i|_k u(d) = F|_{k,\chi} T_K(M)|_k u(d), \end{aligned}$$

where we have used the relation $\chi(M'_i) = \chi(M_i)$, which is valid by Lemma 4.3, and the properties (1.31) of Petersson operators. □

The mapping $d \mapsto |_k u(d)$ defines linear representations of the abelian group $(\mathbb{Z}/q\mathbb{Z})^*$ of invertible elements of the residue ring $\mathbb{Z}/q\mathbb{Z}$ on spaces $\mathfrak{M}_k(K, \chi)$ and $\mathfrak{N}_k(K, \chi)$. Thus, each of the spaces is a direct sum of one-dimensional subspaces invariant under the representation. Each such subspace is spanned by a function F satisfying $F|_k u(d) = \psi(d)F$ for each d coprime to q, where ψ is a character of

$(\mathbb{Z}/q\mathbb{Z})^*$, which can be considered as a Dirichlet character modulo q. Hence, by using Proposition 4.4, we obtain the following.

Proposition 4.5. *Let K be a q-symmetric group of genus n, χ an extendable character of K, and k an integer. Then the following direct sum decompositions hold:*

$$\mathfrak{M}_k(K,\chi)=\bigoplus_{\psi}\mathfrak{M}_k(K,\chi,\psi),\quad \mathfrak{N}_k(K,\chi)=\bigoplus_{\psi}\mathfrak{N}_k(K,\chi,\psi), \tag{4.10}$$

where ψ ranges over Dirichlet characters modulo q, and the ψ-direct summands are defined as subsets of the functions F of the corresponding spaces satisfying the relation $F|_k u(d)=\psi(d)F$ for each d coprime to q; each of the spaces $\mathfrak{M}_k(K,\chi,\psi)$ and $\mathfrak{N}_k(K,\chi,\psi)$ is invariant under all of the Hecke operators $|_{k,\chi}T$ with $T\in\mathcal{L}(K)$.

Diagonalization of Regular Hecke Operators. Here we shall prove, in particular, that under certain conditions, the subspaces of cusp forms are spanned by common eigenfunctions of all regular Hecke operators.

In order to study the action of the Hecke operators on the invariant subspaces introduced above, we shall use the Petersson scalar product (1.70). Note first of all that the regular Hecke operators agree with the Petersson scalar product in the following sense.

Lemma 4.6. *Let K be a q-symmetric group of genus n, χ an extendable character of K, k an integer, and ψ a character modulo q. Suppose that at least one of the modular forms $F, F'\in\mathfrak{M}_k(K,\chi,\psi)$ is a cusp form. Then for each matrix $M\in\Delta(K)$, the scalar product* (1.70) *satisfies the relation*

$$(F|_{k,\chi}T_K(M),F')=\psi(\mu)\overline{\chi}(\mu u(\mu))(F,F'|_{k,\chi}T_M(K)), \tag{4.11}$$

where $\mu=\mu(M)$ is the multiplier of M and $u(\mu)$ is a matrix of the form (4.7).

Proof. By Lemma 3.3, the number of left cosets modulo K contained in the double coset KMK is equal to the index $[K:K\cap M^{-1}KM]$, whereas the number of right cosets is $[K:K\cap MKM^{-1}]$. By Lemma 1.39, these indices are equal. Hence, the number of left cosets modulo K in KMK is equal to the number of right cosets. Since each of the left cosets clearly meets each of the right cosets, it follows that there is a set $\{M_1,\dots,M_\nu\}$ of common representatives for both the left and right cosets modulo K contained in KMK. By using this set of representatives and properties of the scalar product indicated in Theorem 1.28, we get

$$\begin{aligned}(F|_{k,\chi}T_K(M),F')&=\sum_{1\le i\le\nu}(F|_{k,\chi}M_i,F')\\&=\sum_i\overline{\chi}(M_i)(F|_kM_i,F'|_kM_i^{-1}|_kM_i)\\&=\sum_i\overline{\chi}(M_i)\mu(M_i)^{n(k-n-1)}(F,F'|_kM_i^{-1})\end{aligned}$$

$$= \left(F, \sum_i \chi(M_i)\mu(M_i)^{n(k-n-1)} F'|_k M_i^{-1}\right)$$

$$= \left(F, \sum_i \chi(M_i) F'|_k \mu M_i^{-1}\right),$$

since $\mu(M_i) = \mu(M) = \mu$, and we have used the property (1.30) of Petersson operators. Since $F' \in \mathfrak{M}_k(K, \chi, \psi)$, a summand of the last sum can be rewritten in the form

$$\chi(M_i)F'|_k \mu M_i^{-1} = \chi(M_i)F'|_k u(\mu)^{-1} u(\mu)\mu M_i^{-1} = \overline{\psi}(u(\mu))\chi(M_i)F'|_k u(\mu)\mu M_i^{-1}.$$

Let us show now that each of the matrices $u(\mu)\mu M_i^{-1}$ belongs to the double coset KMK. Indeed, since $M_i \in \Delta(K)$, we have $M_i = \gamma N_i$ with $\gamma \in K$ and $N_i \in \Delta(q)$, hence, $u(\mu)\mu M_i^{-1} = u(\mu)\mu N_i^{-1}\gamma^{-1}$, and clearly $u(\mu)\mu N_i^{-1}\gamma^{-1} \in \Delta(q)K = \Delta(K)$. Moreover, by (3.50), this matrix and the matrix M have equal symplectic divisors and so belong to the same double coset modulo K. Since the matrices M_i belong to different right cosets modulo K, it easily follows that the matrices $u(\mu)\mu M_i^{-1}$ form a system of representatives for the left cosets modulo K contained in KMK. Furthermore, since $\chi(u(\mu)\mu M_i^{-1})\chi(M_i) = \chi(u(\mu)\mu M_i^{-1}M_i) = \chi(\mu u(\mu))$, we conclude that $\chi(M_i) = \chi(\mu u(\mu))\overline{\chi}(u(\mu)\mu M_i^{-1})$, whence, finally, we obtain

$$\begin{aligned}\sum_i \chi(M_i)F'|_k \mu M_i^{-1} &= \sum_i \overline{\psi}(u(\mu))\chi(M_i)F'|_k u(\mu)\mu M_i^{-1} \\ &= \overline{\psi}(u(\mu))\chi(\mu u(\mu))\sum_i \overline{\chi}(u(\mu)\mu M_i^{-1})F'|_k u(\mu)\mu M_i^{-1} \\ &= \overline{\psi}(u(\mu))\chi(\mu u(\mu))F'|_{k,\chi} T_K(M).\end{aligned}$$

The formula (4.11) follows. □

Now we are able to prove the following diagonalization theorem.

Theorem 4.7. *Let K be a q-symmetric group of genus n, χ an extendable character of K, k an integer. Then, for each character ψ modulo q, the subspace $\mathfrak{N}_k(K, \chi, \psi)$ of the space of cusp forms $\mathfrak{N}_k(K, \chi)$ has a basis of eigenfunctions, orthogonal with respect to the Petersson product, for all of the Hecke operators $|_{k,\chi}T$ with $T \in \mathcal{L}(K)$. In particular, the whole space $\mathfrak{N}_k(K, \chi)$ has such a basis.*

Proof. By Theorem 1.30, each of the spaces $\mathbf{V} = \mathfrak{N}_k(K, \chi) \subset \mathfrak{M}_k(K, \chi)$ is finite-dimensional over $\mathbb{C}$. Hence the ring of Hecke operators on this space can be presented by matrices of finite order and therefore has finite dimension. Since each Hecke operator is a linear combination of operators $|_{k,\chi}T_K(M)$ with $M \in \Delta(K)$, it follows that there is a finite set $\{M_1, \ldots, M_d\}$ of matrices in Δ such that each Hecke operator on $\mathbf{V}$ is a linear combination of the operators $T_i = |_{k,\chi}T_K(M_i)$. Since, by Theorem 3.45, the ring $\mathcal{L}(K)$ is commutative, the operators $T_1, \ldots, T_d$ commute with each other. Since each linear operator on a nonzero finite-dimensional complex space has a nonzero eigenvector, it easily follows by induction on d that the operators $T_1,\dots,T_d$ on $\mathbf{V}$ have a common eigenfunction $F_1 \neq 0$, provided that $\mathbf{V} \neq \{\mathbf{0}\}$. Let us denote by $\mathbf{V}_1 = \{F \in \mathbf{V} | (F, F_1) = 0\}$ the orthogonal complement to the subspace $\mathbf{V}'_1 = \{\mathbb{C}F_1\}$ in $\mathbf{V}$ relative to the Petersson scalar product. Since, by Theorem 1.38, the scalar product on $\mathbf{V}$ is Hermitian and nondegenerate, it follows that the space $\mathbf{V}$ is the orthogonal direct sum of $\mathbf{V}'_1$ and $\mathbf{V}_1$: $\mathbf{V} = \mathbf{V}'_1 \oplus \mathbf{V}_1$. By formula (4.11) we conclude, then, that the subspace $\mathbf{V}_1$ is also invariant under the operators $T_1,\dots,T_d$, and we may apply the same arguments to the space $\mathbf{V}_1$ in place of $\mathbf{V}$ in order to find a nonzero common eigenfunction $F_2 \in \mathbf{V}_1$ of the operators $T_1,\dots,T_d$, provided that $\mathbf{V}_1 \neq \{\mathbf{0}\}$, and so on. After a finite number of steps we arrive at an orthogonal basis of $\mathbf{V}$ consisting of common eigenfunctions of all operators $T_1,\dots,T_d$ and hence of all regular Hecke operators. □

Exercise 4.8. Show that under the assumptions of Theorem 4.7, the orthogonal complement $\mathfrak{N}_k(K,\chi)^{\perp}$ to the subspace of cusp forms in $\mathfrak{M}_k(K,\chi)$ is invariant relative to all regular Hecke operators.

Exercise 4.9. Show that the Eisenstein series $E_k(z)$ defined in Exercise 1.20 are eigenfunctions for all Hecke operators $|T = T|_{k,1}$ with $T \in \mathcal{L}^1 = D(\Gamma^1, \Delta^1)$. Use this fact to prove that the space $\mathfrak{M}_k^1 = \mathfrak{M}_k(\Gamma^1, 1)$ has a basis of common eigenfunctions for all of the Hecke operators.

Exercise 4.10. Show that the function $\Delta'(z)$ of Exercise 1.36 is a common eigenfunction for all of the Hecke operators on $\mathfrak{M}_k^1$.

4.2 Action of Hecke Operators for $\Gamma_0^n(q)$

In this section we consider in more detail the action of regular Hecke operators on modular forms for the groups $\Gamma_0(q) = \Gamma_0^n(q)$, where n and q are arbitrary positive integers. These groups have not only a distinct arithmetic origin as groups of symmetries of theta series of integral quadratic forms, but also certain technical features that simplify the presentation.

Hecke Operators for $\Gamma_0^n(q)$. We recall that by Lemma 3.44, each group $\Gamma_0(q)$ is q-symmetric with the set $\Delta(\Gamma_0(q)) = \Delta_0(q)$ given by (3.74), and the regular Hecke–Shimura ring of the group is the ring $\mathcal{L}_0(q) = D(\Gamma_0(q), \Delta_0(q))$. The following lemma is a direct consequence of the definitions.

Lemma 4.11. *Let χ be a Dirichlet character modulo q. Then the map*

$$\Gamma_0(q) \ni M = \begin{pmatrix} A & B \\ C & D \end{pmatrix} \mapsto \chi(M) = \chi(\det D) = \overline{\chi}(\det A)$$

is an extendable congruence character of the group $\Gamma_0(q) = \Gamma_0^n(q)$ *with extension on* $\Delta_0(q) = \Delta_0^n(q)$ *given by*

$$\Delta_0(q) \ni M = \begin{pmatrix} A & B \\ C & D \end{pmatrix} \mapsto \chi(M) = \overline{\chi}(\det A). \tag{4.12}$$

We shall fix these extensions of the character χ on $\Gamma_0(q)$ and $\Delta_0(q)$, and introduce the *Petersson operator of weight k and character* χ by

$$\Delta_0(q) \ni M = \begin{pmatrix} A & B \\ C & D \end{pmatrix} : F \mapsto F|_{k,\chi}M = \overline{\chi}(M)F|_k M = \chi(\det A)F|_k M, \tag{4.13}$$

where $|_k$ are the Petersson operators (1.29). In this notation, the Hecke operators (4.5) on spaces $\mathfrak{M}_k(\Gamma_0(q), \chi)$ of modular forms of integral weights k and character χ for $\Gamma_0(q)$ corresponding to elements $T = \sum_i a_i(KM_i) \in \mathcal{L}_0(q)$ take the form

$$F|T = F|_{k,\chi}T = \sum_i a_i F|_{k,\chi}M_i. \tag{4.14}$$

We shall refer to these operators as *regular Hecke operators*.

Operating on modular forms, regular Hecke operators operate on the Fourier coefficients of the modular forms. According to Theorem 1.23, each modular form $F \in \mathfrak{M}_k(\Gamma_0(q), \chi)$ has a Fourier expansion

$$F(Z) = \sum_{A \in \mathbb{E}, A \geq 0} f(A)e^{\pi i \sigma(AZ)} \tag{4.15}$$

with constant Fourier coefficients $f(A)$ satisfying

$$f({}^tVAV) = (\det V)^k \chi(\det V) f(A) \quad \text{for all } A \in \mathbb{E} \text{ and } V \in \Lambda^n = \mathrm{GL}_n(\mathbb{Z}), \tag{4.16}$$

where $\mathbb{E} = \mathbb{E}^n$. For $T \in \mathcal{L}_0(q)$ we shall denote by $(f|T)(A) = (f|_{k,\chi}T)(A)$ with $A \in \mathbb{E}$ the Fourier coefficients of the modular form $F|T$, so that

$$F|T = \sum_{A \in \mathbb{E},\, A \geq 0} (f|T)(A)e^{\pi i \sigma(AZ)}. \tag{4.17}$$

The action of the Hecke operators on the Fourier coefficients of the modular forms of the space $\mathfrak{M}_k(\Gamma_0(q), \chi)$ considered as functions on $\mathbb{E}$ satisfying (4.15) obviously satisfies the rules

$$f|aT + bT' = af|T + bf|T', \qquad f|TT' = (f|T)|T' \qquad (T, T' \in \mathcal{L}_0(q)) \tag{4.18}$$

and so defines a linear representation of the ring $\mathcal{L}_0(q)$ on the functions.

Hecke Operators for Γ_0^n. In Section 3.5 we considered decompositions of elements of rings $\mathcal{L}_0(q)$ into components belonging to rings of double cosets of triangular groups $\Gamma_0 = \Gamma_0^n$. In general, Hecke operators corresponding to double cosets of the triangular groups do not map spaces $\mathfrak{M}_k(\Gamma_0(q), \chi)$ into themselves. In order to apply the decompositions, we have first to extend the spaces of modular forms. To this end, we introduce the spaces $\mathfrak{F} = \mathfrak{F}^n$ of all Fourier series with constant Fourier coefficients of the form

$$G(Z) = \sum_{A \in \mathbb{E}^n,\, A \geq 0} g(A) e^{\pi i \sigma(AZ)},$$

which converge absolutely on $\mathbb{H} = \mathbb{H}^n$ and uniformly on subsets (1.37). For a character ε of the group $\{\pm 1\}$, we define the subspace $\mathfrak{F}_\varepsilon = \mathfrak{F}_\varepsilon^n$ of $\mathfrak{F}$ consisting of the functions F satisfying

$$F((\widehat{V}Z + U)V^{-1}) = \varepsilon(\det V) F \quad \text{for each } \begin{pmatrix} \widehat{V} & U \\ 0 & V \end{pmatrix} \in \Gamma_0, \tag{4.19}$$

where $\widehat{V} = {}^tV^{-1}$. These conditions are obviously equivalent to the conditions that F belongs to $\mathfrak{F}$ and that its Fourier coefficients satisfy

$$f({}^tVAV) = \varepsilon(\det V) f(A) \quad \text{for all } A \in \mathbb{E} = \mathbb{E}^n \text{ and } V \in \Lambda = \Lambda^n. \tag{4.20}$$

It is clear that $\mathfrak{F}_\varepsilon = \{0\}$ if $\varepsilon((-1)^n) = -1$. We define the action of the Hecke operators of the Hecke–Shimura rings $\boldsymbol{\mathcal{L}}(q)$ of the form (3.118) on these spaces according to the general scheme with the help of the Petersson operators (4.13).

Lemma 4.12. *Let k be an integer, q a positive integer, and χ a Dirichlet character modulo q. Let ε be the character of the group $\{\pm 1\}$ satisfying $\varepsilon(-1) = (-1)^k \chi(-1)$. Then for $\mathbf{T} = \sum_i a_i(\Gamma_0 M_i) \in \boldsymbol{\mathcal{L}}_0(q)$, the operator*

$$\mathfrak{F}_\varepsilon \ni F \mapsto F|_{k,\chi}\mathbf{T} = \sum_i a_i F|_{k,\chi} M_i \tag{4.21}$$

is independent of the choice of representatives $M_i \in \Gamma_0 M_i$ and maps the space $\mathfrak{F}_\varepsilon$ into itself; the mapping $\mathbf{T} \mapsto |_{k,\chi}\mathbf{T}$ is a linear representation of the ring $\boldsymbol{\mathcal{L}}(q)$ on $\mathfrak{F}_\varepsilon$ compatible with the standard representation of $\mathcal{L}_0(q)$ on $\mathfrak{M}_k(\Gamma_0(q), \chi)$ and the embeddings $\mathfrak{M}_k(\Gamma_0(q), \chi) \subset \mathfrak{F}_\varepsilon$ and $\iota : \mathcal{L}_0(q) \mapsto \mathbf{L}(q) \subset \boldsymbol{\mathcal{L}}(q)$.

Proof. By definition and the assumption on ε, the subspace of Γ_0-invariant functions of $\mathfrak{F}$ consists of functions satisfying

$$F \stackrel{!}{=} F|_{k,\chi} \begin{pmatrix} \widehat{V} & U \\ 0 & V \end{pmatrix} = \overline{\chi}(\det V)(\det V)^{-k} F\big((\widehat{V}X + U)V^{-1}\big)$$
$$= \varepsilon(\det V)^{-1} F\big((\widehat{V}X + U)V^{-1}\big) \quad \text{for each } \begin{pmatrix} \widehat{V} & U \\ 0 & V \end{pmatrix} \in \Gamma_0,$$

i.e., it coincides with $\mathfrak{F}_\varepsilon$, and the first part of the lemma follows from Lemma 4.1. As to compatibility of the representations with triangular embeddings, for $F \in \mathfrak{M}_k(\Gamma_0(q), \chi)$ and $T = \sum_i a_i(\Gamma_0(q) M_i) \in \mathcal{L}_0(q)$ with triangular representatives M_i, we have $\iota(T) = \sum_i a_i(\Gamma_0 M_i) \in \boldsymbol{\mathcal{L}}(q)$ and $F|_{k,\chi} T = \sum_i a_i F|_{k,\chi} M_i = F|_{k,\chi}\iota(T)$. □

The following lemma gives formulas for the action of the Hecke operators on the Fourier coefficients of the functions belonging to $\mathfrak{F}_\varepsilon$.

Lemma 4.13. *In the notation of the previous lemma, the Fourier coefficients of the image of a function $F \in \mathfrak{F}_\varepsilon$ with Fourier coefficients $f(A)$ under the action of the Hecke operator corresponding to a double coset $\mathbf{T}(M) \in \mathcal{L}(q)$ introduced in Lemma* 3.58 *can be written in the form*

$$\begin{aligned}(f|_{k,\chi}\mathbf{T}(M))(A) \\ = \mu^{nk-\frac{n(n+1)}{2}}\chi(\mu)^n \sum_{D'\in\Lambda\backslash\Lambda D\Lambda, A[{}^tD']\in\mu\mathbb{E}} \overline{\chi}(\det D')(\det D')^{-k} \\ \times s(\mu^{-1}A[{}^tD'], D', M) f(\mu^{-1}A[{}^tD']) \\ \left(M = \begin{pmatrix}\mu\widehat{D} & B\\ 0 & D\end{pmatrix} \in \Delta_0 \cap \Delta_{\langle q\rangle}, A \in \mathbb{E}, A \geq 0\right),\end{aligned} \tag{4.22}$$

where $\mu = \mu(M)$, and we use the trigonometric sums

$$s(Q, D', M) = \sum_{B'\in B(D',\,M)/\ mod\ D'} e^{\pi i \sigma(QB'(D')^{-1})} \qquad (Q \in \mathbb{E}). \tag{4.23}$$

Proof. The formula (4.22) is an immediate consequence of the definition of Hecke operators and Lemma 3.58. □

The next lemma contains a specialization of the previous formulas in cases of certain important operators.

Lemma 4.14. *In the above notation, for each d coprime to q, the following formulas hold for the action on Fourier coefficients of Hecke operators corresponding to elements*

$$\Pi_-(d) = \Pi_-^n(d) = \mathbf{T}\left(\begin{pmatrix} d\cdot 1_n & 0\\ 0 & 1_n\end{pmatrix}\right), \quad \Pi_+(d) = \Pi_+^n(d) = \mathbf{T}\left(\begin{pmatrix} 1_n & 0\\ 0 & d\cdot 1_n\end{pmatrix}\right),$$

$\langle\mathbf{d}\rangle = \langle\mathbf{d}\rangle_n = \mathbf{T}(d\cdot 1_{2n}) \in \mathcal{L}^n(q)$, and the element $\Pi(d) = \mathbf{T}(\mathrm{diag}(d,1,1,d)) \in \mathcal{L}^2(q)$ introduced in Section 3.5*:*

$$(f|_{k,\chi}\Pi_-(d))(A) = \chi(d)^n d^{nk-\frac{n(n+1)}{2}} \times \begin{cases} f(d^{-1}A) & \text{if } A \in d\mathbb{E},\\ 0 & \text{if } A \notin d\mathbb{E};\end{cases} \tag{4.24}$$

$$(f|_{k,\chi}\Pi_+(d))(A) = f(dA); \tag{4.25}$$

$$(f|_{k,\chi}\langle\mathbf{d}\rangle) = \chi(d)^n d^{nk-n(n+1)} f(A); \tag{4.26}$$

$$(f|_{k,\chi}\Pi(d))(A) = \chi(d)d^{k-2} \sum_{\substack{D\in\Lambda_+\backslash\Lambda_+\mathrm{diag}(1,d)\Lambda_+,\\ A[{}^tD]\in d\mathbb{E}^2}} f(d^{-1}A[{}^tD]), \tag{4.27}$$

where $\Lambda_+ = \Lambda_+^2 = \mathrm{SL}_2(\mathbb{Z})$.

Proof. It follows from Lemma 3.60 and the definitions that for matrices

$$M = \begin{pmatrix} A & B \\ 0 & D \end{pmatrix} = \begin{pmatrix} d\cdot 1_n & 0 \\ 0 & 1_n \end{pmatrix}, \begin{pmatrix} 1_n & 0 \\ 0 & d\cdot 1_n \end{pmatrix}, \text{ and } \begin{pmatrix} d\cdot 1_n & 0 \\ 0 & d\cdot 1_n \end{pmatrix},$$

one can take the sets $\{D'\} = \{1_n\}$, $\{d\cdot 1_n\}$, and $\{d\cdot 1_n\}$ as sets of representatives of left cosets $\Lambda\backslash\Lambda D\Lambda$ and take the sets

$$\{0 = 0_n\}, \{B = {}^tB = (b_{ij}) | 0 \le b_{ij} < d, (1 \le i \le j \le n)\}, \text{ and } \{0 = 0_n\}$$

as sets of representatives for $B(D',M)/\bmod D'$, respectively. It clearly follows that the trigonometric sum (4.23) takes the values $s(Q,D',M) = 1$ for each $Q \in \mathbb{E}^n$ in the first and third cases and the values

$$s(Q,D',M) = \begin{cases} d^{n(n+1)/2} & \text{if } Q \in d\mathbb{E}, \\ 0 & \text{if } Q \notin d\mathbb{E}, \end{cases}$$

in the second case. With these remarks, the formulas (4.24)–(4.26) follow directly from formula (4.22). Similarly, since clearly, $\Lambda\backslash\Lambda \operatorname{diag}(1,d)\Lambda = \Lambda_+\backslash\Lambda_+\operatorname{diag}(1,d)\Lambda_+$, by formulas (3.134) and (4.22), we obtain

$$\begin{aligned}(f|_{k,\chi}\Pi(d))(A) \\ = d^{k-3}\chi(d) \sum_{\substack{D\in\Lambda_+\backslash\Lambda_+\operatorname{diag}(1,d)\Lambda_+,\\ A[{}^tD]\in d\mathbb{E}^2}} s(d^{-1}A[{}^tD], D, \operatorname{diag}(1,d)) f(d^{-1}A[{}^tD]),\end{aligned}$$

where, since in this case $B(D,\operatorname{diag}(1,d)) = B(D)$ is the set (3.58), we can write

$$s(d^{-1}A[{}^tD], D, \operatorname{diag}(1,d)) = \sum_{B\in B(D)/\bmod D} e^{\pi i\sigma(d^{-1}A[{}^tD]BD^{-1})}.$$

Thus, in order to prove the formula (4.27), it is sufficient to show that the last sum equals d for every matrix $A \in \mathbb{E}^2$ satisfying $d^{-1}A[{}^tD] \in \mathbb{E}^2$. According to Lemma 3.3, each representative $D \in \Lambda_+\backslash\Lambda_+D_0\Lambda_+$, where $D_0 = \operatorname{diag}(1,d)$, can be taken in the form $D = D_0\lambda$ with $\lambda \in \Lambda_+$. It follows then from the definitions that $B(D) = B(D_0\lambda) = B(D_0)\lambda$, and one can take $B(D)/\bmod D = \{B(D_0)/\bmod D_0\}\lambda$ and set $B(D_0)/\bmod D_0 = \{B = \left(\begin{smallmatrix} 0 & 0 \\ 0 & b\end{smallmatrix}\right) | 0 \le b < d\}$. Hence, the last sum is equal to

$$\sum_{0\le b<d} e^{\pi i\sigma\left(d^{-1}(\lambda A{}^t\lambda)[D_0]\begin{pmatrix} 0 & 0 \\ 0 & b/d\end{pmatrix}\right)} = \sum_{0\le b<d} e^{\pi i(\lambda A{}^t\lambda)_{22}b} = d. \quad \square$$

Exercise 4.15. For $Q \in \mathbb{E}^n$ and $D \in \Sigma^n$, set

$$s(Q,D) = \sum_{B\in B(D)/\bmod D} e^{\pi i\sigma(QBD^{-1})}$$

and $\mathbb{E}^n\{D\} = \{Q \in \mathbb{E}^n | s(Q, D) \neq 0\}$. Prove the following assertions:

(1) If $\lambda, \lambda' \in \Lambda^n$, then $s(Q, \lambda D \lambda') = s(Q[\widehat{\lambda}], D)$ and $\mathbb{E}^n\{\lambda D \lambda'\} = \lambda \mathbb{E}^n\{D\}$;
(2) If $\mathrm{ed}(D) = \mathrm{diag}(d_1, \ldots, d_n)$, then

$$s(Q, D) = \begin{cases} d_1^n d_2^{n-1} \cdots d_n & \text{if } Q \in \mathbb{E}^n\{D\}, \\ 0 & \text{if } Q \notin \mathbb{E}^n\{D\}; \end{cases}$$

(3) If $D = \mathrm{ed}(D) = \mathrm{diag}(d_1, \ldots, d_n)$ and $Q = \big((1 + e_{ij}) q_{ij}\big)$ with $(e_{ij}) = 1_n$ and $q_{ij} \in \mathbb{Z}$, then

$$Q \in \mathbb{E}^n\{D\} \text{ if and only if } \quad q_{ij} \equiv 0 \pmod{d_i} \text{ for } 1 \le i \le j \le n.$$

Exercise 4.16. In the notation of Lemma 4.14 and Exercise 3.62, prove, for $i = 0, 1, \ldots, n$ and $A \in \mathbb{E}^n$, the formula

$$(f|_{k,\chi}\Pi_i^n(d))(A) = d^{(n-i)k - \frac{n(n+1)}{2} + \frac{i(i+1)}{2}} \chi(d)^{n-i} \sum_{\substack{D \in \Lambda_+ \backslash \Lambda_+ D_i^n(d) \Lambda_+, \\ A[{}^tD] \in d\mathbb{E}}} f(d^{-1} A[{}^tD]),$$

where $\Lambda_+ = \mathrm{SL}_n(\mathbb{Z})$.

Exercise 4.17. In the notation of Lemma 4.14 and (3.124), for $A \in \mathbb{E}^n$ and $a \in \mathbb{N}$ coprime to q, prove the formula

$$(f|_{k,\chi}\mathbf{T}(m))(A) = m^{nk - \frac{n(n+1)}{2}} \chi(m)^n \sum_{\substack{d_1, \ldots, d_n \in \mathbb{N}, \\ d_1 | \cdots | d_n | m}} \overline{\chi}(d_1 \cdots d_n) d_1^{n-k} d_2^{n-1-k} \cdots d_n^{1-k}$$
$$\times \sum_{\substack{D \in \Lambda_+ \backslash \Lambda_+ \mathrm{diag}(d_1, \ldots, d_n) \Lambda_+, \\ A[{}^tD] \in m\mathbb{E}}} f(m^{-1} A[{}^tD]).$$

4.3 Hecke Operators and Siegel Operator

In this section we shall show that Hecke operators on the spaces $\mathfrak{M}_k(\Gamma_0^n(q), \chi)$ and $\mathfrak{F}_\varepsilon^n$ are consistent with the Siegel operator Φ defined in Section 1.3 in the sense that there is a homomorphism $T \mapsto T'$ of the Hecke–Shimura rings from genus n to genus $n-1$ such that $|_{k,\chi} T | \Phi = |\Phi|_{k,\chi} T'$. Since the global rings of operators under consideration are generated by local subrings, we shall restrict ourselves to a study of the local situation.

Action of the Siegel Operator on Fourier Series. First of all, we note that for $K = \Gamma_0^n(q)$ and a character $\chi = \chi^n$ of the form (4.12) associated with a Dirichlet character χ modulo q, the group $\overleftarrow{K}$ defined by (1.57) is clearly $\Gamma_0^{n-1}(q)$, and the character $\overleftarrow{\chi}$ of this group is the character $\chi = \chi^{n-1}$ of the form (4.12) associated with the same Dirichlet character χ modulo q. Next, we note that the conditions (1.55) and (1.56) in fact define the Siegel operator Φ on the whole space of Fourier series $\mathfrak{F}^n$ with $n \geq 1$ and map the space into $\mathfrak{F}^{n-1}$ if we set $\mathfrak{F}^0 = \mathbb{C}$. With these conventions we have the following extension of Theorem 1.28.

Proposition 4.18. *Let k be an integer, q a positive integer, and χ a Dirichlet character modulo q. Let ε be the character of the group $\{\pm 1\}$ satisfying $\varepsilon(-1) = (-1)^k\chi(-1)$. Then for $n \geq 1$, the Siegel operator Φ maps the space $\mathfrak{F}^n_\varepsilon$ into $\mathfrak{F}^{n-1}_\varepsilon$, where we set $\mathfrak{F}^0_\varepsilon = \mathbb{C}$ if $\varepsilon(-1) = 1$, and $\mathfrak{F}^0_\varepsilon = \{0\}$ if $\varepsilon(-1) = -1$; in addition, the subspace $\mathfrak{M}_k(\Gamma^n_0(q), \chi) \subset \mathfrak{F}^n_\varepsilon$ is mapped into $\mathfrak{M}_k(\Gamma^{n-1}_0(q), \chi)$.*

Proof. If $F \in \mathfrak{F}^1_\varepsilon$, then $F|\Phi$ is just the constant term of the Fourier series F. This term equals zero when $\varepsilon(-1) = -1$, because by (4.19), we have $F((-1)Z(-1)) = F(Z) = \varepsilon(-1)F(Z)$. If $F \in \mathfrak{F}^n_\varepsilon$ with $n > 1$, $\begin{pmatrix} \widehat{V}' & U' \\ 0 & V' \end{pmatrix} \in \Gamma^{n-1}_0$, and $V = \begin{pmatrix} V' & 0 \\ 0 & 1 \end{pmatrix}$, $U = \begin{pmatrix} U' & 0 \\ 0 & 0 \end{pmatrix}$, then $\begin{pmatrix} \widehat{V} & U \\ 0 & V \end{pmatrix} \in \Gamma^n_0$, and we have $F((\widehat{V}Z + U)V^{-1}) = \varepsilon(\det V)F = \varepsilon(\det V')F$. Taking here $Z = \begin{pmatrix} Z' & 0 \\ 0 & iy \end{pmatrix}$ with $Z' \in \mathbb{H}^{n-1}$ and $y > 0$ and passing to the limit as $y \to +\infty$, we get the relation $(F|\Phi)((\widehat{V}'Z' + U')(V')^{-1}) = \varepsilon(\det V')F|\Phi$. The rest follows from Theorem 1.28. □

Let us look now at the relation of the Siegel operator to the Hecke operators $|_{k,\chi}\mathbf{T}$ with an integral k and a Dirichlet character χ modulo q corresponding to elements of local Hecke–Shimura rings $\mathcal{L}^n_p = \mathcal{L}^n_p(q)$ for primes p not dividing q. Since the values of χ are, generally speaking, not rational, it is convenient to consider linear combinations of double cosets contained in these rings, not only with integral or rational coefficients as before, but also with arbitrary complex coefficients, i.e., elements of the complexifications $\acute{\mathcal{L}}^n_p = D_{\mathbb{C}}(\Gamma^n_0, \Delta^n_0 \cap \Delta^n_p)$ of the rings $\mathcal{L}^n_p$. Let

$$F(Z) = \sum_{A \in \mathbb{E}, A \geq 0} f(A)e^{\pi i \sigma(AZ)} \in \mathfrak{F}^n_\varepsilon \tag{4.28}$$

be a Fourier series of genus $n \geq 1$ and character ε, where $\varepsilon(-1) = (-1)^k\chi(-1)$, and let

$$\mathbf{T} = \sum_i a_i \left(\Gamma^n_0 \begin{pmatrix} p^{\delta_i}\widehat{D}_i & B_i \\ 0 & D_i \end{pmatrix}\right) \in \acute{\mathcal{L}}^n_p \tag{4.29}$$

be $\Gamma_0 = \Gamma^n_0$-invariant linear combination of left cosets with complex coefficients. By (4.14), we have

$$\begin{aligned} F|_{k,\chi}\mathbf{T} &= \sum_i a_i \left(\sum_{A \in \mathbb{E},\, A \geq 0} f(A)e^{\pi i \sigma(AZ)} \right) |_{k,\chi} \begin{pmatrix} p^{\delta_i}\widehat{D}_i & B_i \\ 0 & D_i \end{pmatrix} \\ &= \sum_{i,\, A \in \mathbb{E},\, A \geq 0} a_i p^{\delta_i(nk - \frac{n(n+1)}{2})} \chi(\det(p^{\delta_i}\widehat{D}_i))(\det D_i)^{-k} \\ &\quad \times f(A)e^{\pi i \sigma(p^{\delta_i}A[\widehat{D}_i]Z + AB_iD_i^{-1})}. \end{aligned}$$

Since every matrix of the form $\begin{pmatrix} \widehat{\lambda} & 0 \\ 0 & \lambda \end{pmatrix}$ with $\lambda \in \Lambda = \Lambda^n$ is contained in Γ_0, by Lemma 3.18 we conclude that each of the matrices D_i in the last sum can be taken in the form $D_i = \begin{pmatrix} D_i' & * \\ 0 & p^{d_i} \end{pmatrix}$. For such D_i and $A = \begin{pmatrix} A' & * \\ * & a \end{pmatrix}$, the matrix $p^{\delta_i} A[\widehat{D}_i]$ has the form $\begin{pmatrix} * & * \\ * & p^{\delta_i - 2d_i} a \end{pmatrix}$. Hence, when we set $Z = \begin{pmatrix} Z' & 0 \\ 0 & \sqrt{-1}y \end{pmatrix}$ with $Z' \in \mathbb{H}^{n-1}$ and $y > 0$ and turn y to $+\infty$, each summand of the last sum corresponding to a matrix A with $a = A_{nn} > 0$ tends to zero. As to other summands, since $A \geq 0$, it follows that $A = \begin{pmatrix} A' & 0 \\ 0 & 0 \end{pmatrix}$. Further, for such A, Z, D_i, and $B_i = \begin{pmatrix} B_i' & * \\ * & * \end{pmatrix}$, we have

$$p^{\delta_i} A[\widehat{D}_i] = p^{\delta_i} \begin{pmatrix} A'[\widehat{D}_i'] & 0 \\ 0 & 0 \end{pmatrix} \quad \text{and} \quad AB_i D_i^{-1} = \begin{pmatrix} A'B_i'(D_i')^{-1} & 0 \\ 0 & 0 \end{pmatrix}.$$

Using these relations, since the series F converges uniformly on the sets $\mathbb{H}^n_\varepsilon$, we obtain

$$\begin{aligned}
(F|_{k,\,\chi}\mathbf{T})|\Phi &= \lim_{y \to +\infty} (F|_{k,\,\chi}\mathbf{T}) \begin{pmatrix} Z' & 0 \\ 0 & \sqrt{-1}y \end{pmatrix} \\
&= \sum_{i,\, A' \in \mathbb{E}^{n-1},\, A' \geq 0} a_i p^{\delta_i (nk - \frac{n(n+1)}{2})} \chi(p^{\delta_i - d_i} \det(p^{\delta_i} \widehat{D}_i'))(p^{d_i} \det D_i')^{-k} \\
&\quad \times f\left(\begin{pmatrix} A' & 0 \\ 0 & 0 \end{pmatrix} \right) e^{\pi i \sigma (p^{\delta_i} A'[\widehat{D}_i'] Z' + A'B_i'(D_i')^{-1})} \\
&= (F|\Phi)|_{k,\chi} \sum_i a_i p^{\delta_i (k-n)} \chi(p^{\delta_i - d_i}) p^{-kd_i} \begin{pmatrix} p^{\delta_i} \widehat{D}_i' & B_i' \\ 0 & D_i' \end{pmatrix} \\
&= (F|\Phi)|_{k,\chi} \Psi(\mathbf{T}),
\end{aligned}$$

where for $\mathbf{T}$ of the form (4.29) with $D_i = \begin{pmatrix} D_i' & * \\ 0 & p^{d_i} \end{pmatrix}$ and $B_i = \begin{pmatrix} B_i' & * \\ * & * \end{pmatrix}$, we set

$$\begin{aligned}
\Psi(\mathbf{T}) &= \Psi^n_{k,\chi}(\mathbf{T}) \\
&= \sum_i a_i (\chi(p) p^{k-n})^{\delta_i} (\overline{\chi}(p) p^{-k})^{d_i} \left(\Gamma_0^{n-1} \begin{pmatrix} p^{\delta_i} \widehat{D}_i' & B_i' \\ 0 & D_i' \end{pmatrix} \right).
\end{aligned} \tag{4.30}$$

If $\gamma' = \begin{pmatrix} \alpha' & \beta' \\ 0 & \delta' \end{pmatrix} \in \Gamma_0^{n-1}$, then the matrix $\gamma = \overrightarrow{\gamma}'$ of the form (1.57) belongs to the group Γ_0^n, and the equality $\mathbf{T}\gamma = \mathbf{T}$ implies the relation $\Psi(\mathbf{T})\gamma' = \Psi(\mathbf{T})$. Hence $\Psi(\mathbf{T}) \in \acute{\mathcal{L}}_p^{n-1}$, where we set $\acute{\mathcal{L}}_p^0 = \mathbb{C}$. It is readily verified that the mapping $\mathbf{T} \mapsto \Psi(\mathbf{T})$ is a homomorphism of the $\mathbb{C}$-algebras. Thus, we come to the following theorem.

Zharkovskaya Commutation Relations.

Theorem 4.19. (*Zharkovskaya commutation relations*). *Let k be an integer, q a positive integer, χ a Dirichlet character modulo q, and ε the character of the group $\{\pm 1\}$ satisfying $\varepsilon(-1) = (-1)^k \chi(-1)$. Then for every Fourier series $F \in \mathfrak{F}^n_\varepsilon$ of* *genus $n \geq 1$ and element $\mathbf{T} \in \acute{\mathcal{L}}_p^n$ of the form (4.29) with a prime p not dividing q, the following relation holds:*

$$(F|_{k,\chi}\mathbf{T})|\Phi = (F|\Phi)|_{k,\chi}\Psi(\mathbf{T}), \tag{4.31}$$

where $\Phi : \mathfrak{F}_\varepsilon^n \mapsto \mathfrak{F}_\varepsilon^{n-1}$ is the Siegel operator, and $\Psi(\mathbf{T}) \in \acute{\mathcal{L}}_p^{n-1}$ is the element (4.30); the mapping $\Psi : \acute{\mathcal{L}}_p^n \mapsto \acute{\mathcal{L}}_p^{n-1}$ is a $\mathbb{C}$-linear ring homomorphism; if $\mathbf{T} = \mathbf{T}(M) \in \mathcal{L}_p^n$ with $n > 1$ is an element of the form (3.19), and $\Omega(\mathbf{T}) = \Omega_p^n(\mathbf{T})(x_0, x_1, \dots, x_n)$ is the spherical polynomial (3.140), then

$$\begin{aligned}&\Omega_p^{n-1}(\Psi(\mathbf{T}(M)))(x_0, x_1, \dots, x_{n-1})\\ &= \Omega_p^n(\mathbf{T}(M))(\chi(p)p^{k-n}x_0, x_1, \dots, x_{n-1}, \overline{\chi}(p)p^{n-k}),\end{aligned} \tag{4.32}$$

where we assume that the spherical mapping is extended to complexifications of Hecke–Shimura rings by linearity.

Proof. All assertions of the theorem, except for the formula (4.32), have already been proved. If $T(M) = \sum_i \left(\Gamma_0^n \begin{pmatrix} p^\delta \widehat{D}_i & B_i \\ 0 & D_i \end{pmatrix}\right)$, then by formula (4.30) and the definition of spherical mapping (see (3.140) and (3.43)), we obtain

$$\begin{aligned}\Omega_p^{n-1}(\Psi(\mathbf{T}(M))) &= \sum_i (\chi(p)p^{(k-n)})^\delta (\overline{\chi}(p)p^{-k})^{d_i} x_0^\delta \omega_p^{n-1}(\Lambda^{n-1}D_i')(x_1, \dots, x_{n-1})\\ &= \sum_i (x_0\chi(p)p^{k-n})^\delta \omega_p^n\left(\Lambda^n \begin{pmatrix} D_i' & * \\ 0 & p^{d_i}\end{pmatrix}\right)(x_1, \dots, x_{n-1}, \overline{\chi}(p)p^{n-k}),\end{aligned}$$

which proves the formula (4.32). □

Let us look now at the restriction of the mapping Ψ to the complexification $\acute{\mathbf{L}}_p$ of the isomorphic image (3.123) in $\mathcal{L}_p$ of the local symplectic Hecke–Shimura ring $\mathcal{L}_p(\Gamma_0(q))$.

Proposition 4.20. *In the notation and under the assumption of Theorem 4.19, the mapping $\Psi = \Psi_{k,\chi}^n$ with $n > 1$ transforms the complexification $\acute{\mathbf{L}}_p^n$ into the ring $\acute{\mathbf{L}}_p^{n-1}$; the mapping*

$$\Psi = \Psi_{k,\chi}^n : \acute{\mathbf{L}}_p^n \mapsto \acute{\mathbf{L}}_p^{n-1} \tag{4.33}$$

is epimorphic, except for the case $n = k$ and $\chi(p) = -1$, in which case the image $\Psi(\acute{\mathbf{L}}_p^n)$ is a proper subring of $\acute{\mathbf{L}}_p^{n-1}$.

Proof. It follows from Theorem 3.51 that the extended spherical mapping $\Omega = \Omega_p^n : \acute{\mathbf{L}}_p^n \mapsto \mathbb{C}[x_0, x_1, \dots, x_n]$ is an isomorphism of the ring $\acute{\mathbf{L}}_p^n$ onto the subring $\mathbb{C}[x_0, x_1, \dots, x_n]_W$ of $W = W^n$-invariant polynomials. Relations (4.32) imply that the spherical polynomial $\Omega_p^{n-1}(\Psi(\mathbf{T}))$ of an element $\mathbf{T} \in \acute{\mathcal{L}}_p^n$ is W^{n-1}–invariant if $\Omega_p^n(\mathbf{T})$ is W^n-invariant. Thus, $\Psi(\acute{\mathbf{L}}_p^n) \subset \acute{\mathbf{L}}_p^{n-1}$.

The ring $\Psi(\acute{\mathbf{L}}_p^n)$ is generated over $\mathbb{C}$ by Ψ-images of a system of generators of the ring $\acute{\mathbf{L}}_p^n$ or the ring $\mathcal{L}_p^n$. The relations (4.32) allow one to find spherical polynomials corresponding to the images if spherical polynomials for the generators are known. Computation of spherical polynomials corresponding to generators for an arbitrary genus $n > 1$ is not hard, but it requires a lot of space. In order to illustrate the computations, we restrict ourselves here to the simplest case $n = 2$. It was shown in the course of the proof of Theorem 3.51 that the sets $\{x_0^2x_1, x_0(1+x_1)\}$ and $\{x_0^2x_1x_2, x_0(1+x_1)(1+x_2), x_0^2(x_1+x_2)(1+x_1x_2)\}$ consist of spherical polynomials corresponding to certain systems of generators (over $\mathbb{Q}$) of the rings $\mathcal{L}_p^1$ and $\mathcal{L}_p^2$, respectively. It follows from (4.32) that the image $\Psi(\acute{\mathbf{L}}_p^2)$ is generated over $\mathbb{C}$ by the elements whose spherical polynomials are equal to the polynomials $\chi(p)p^{k-2}x_0^2x_1$, $\chi(p)p^{k-2}x_0(1+x_1)(1+\overline{\chi}(p)p^{2-k})$, and $\chi(p)^2p^{2(k-2)}x_0^2(x_1 + \overline{\chi}(p)p^{2-k})(1+\overline{\chi}(p)p^{2-k}x_1)$. If $1+\overline{\chi}(p)p^{2-k} \neq 0$, then clearly these elements generate the whole ring $\acute{\mathbf{L}}_p^1$. But if $1+\overline{\chi}(p)p^{2-k} = 0$, i.e., $k = 2$ and $\overline{\chi}(p) = \chi(p) = -1$, then the image is generated by the elements with spherical polynomials $x_0^2x_1$ and $x_0^2(1-x_1)^2 = \left(x_0(1+x_1)\right)^2 - 4x_0^2x_1$ and so cannot be equal to $\acute{\mathbf{L}}_p^1$. □

Diagonalization of Hecke Operators for the Full Modular Group. The Zharkovskaya relations allow one to reduce certain problems relating to Hecke operators on modular forms of a given genus to similar problems for cusp forms and modular forms of smaller genera. In particular, the problem of simultaneous diagonalization of Hecke operators on invariant spaces of modular forms is a problem of this kind. Theorem 4.7 solves the problem of diagonalization for cusp forms. Here we shall consider the problem for invariant subspaces of the space

$$\mathfrak{M}_k^n = \mathfrak{M}_k(\Gamma^n) = \mathfrak{M}_k(\Gamma^n, 1) \tag{4.34}$$

of modular forms of integral nonnegative weight k for the full modular group Γ^n with the trivial character, but first we shall prove a useful lemma on cusp forms contained in these spaces.

Lemma 4.21. *Let*

$$F = \sum_{A \in \mathbb{E}^n,\, A \geq 0} f(A)e^{\pi i \sigma(AZ)} \in \mathfrak{M}_k^n \tag{4.35}$$

be a modular form of weight k for the full modular group Γ^n *then the following three conditions are equivalent:*

(a) F is a cusp form.
(b) $F|\Phi = 0$, *where* Φ *is the Siegel operator* (1.56).
(c) The Fourier coefficients of F satisfy the rule $f(A) = 0$ *if* $\det A = 0$.

Proof. Condition (a) implies (b), by Lemma 1.29. According to formula (1.54), condition (b) implies (c). Let $M \in \mathbb{G}^n(\mathbb{Q})$ be a symplectic matrix of genus n with rational entries and positive multiplier $\mu = \mu(M)$. It follows from Proposition 3.35(1) that there is a matrix $\gamma \in \Gamma^n$ such that $M = \gamma M'$ with triangular $M' = \begin{pmatrix} A' & B' \\ 0 & D' \end{pmatrix}$. Then, if the condition (c) is fulfilled, we have

$$\begin{aligned} F|_k M &= F|_k \gamma M' = F|_k \gamma|_k M' = F|_k M' \\ &= \mu^{nk-\frac{n(n+1)}{2}} (\det D')^{-k} \sum_{A \in \mathbb{E}^n,\ A>0} f(A) e^{\pi i \sigma(A(A'Z+B')(D')^{-1})} \\ &= \mu^{nk-\frac{n(n+1)}{2}} (\det D')^{-k} \sum_{A \in \mathbb{E}^n,\ A>0} f(A) e^{\pi i \sigma(AB'(D')^{-1}} e^{\pi i \sigma((D')^{-1}AA'Z)}. \end{aligned}$$

Each of the matrices $(D')^{-1}AA' = \mu^{-1}\,{}^tA'AA'$ is symmetric and positive definite together with A, whence $F|_k M|\Phi = 0$, by the definition of Φ. Thus, by Lemma 1.29, F is a cusp form. □

Theorem 4.22. *Every subspace* $\mathbf{V} \subset \mathfrak{M}_k^n$ *invariant with respect to all Hecke operators* $|_k T = |_{k,1} T$ *with* $T \in \mathcal{L}^n = \mathcal{L}(\Gamma^n)$ *has a basis of common eigenfunctions for all of the operators.*

Proof. The same arguments that were used to prove Theorem 4.7 show that the theorem is true for invariant subspaces of the space $\mathfrak{N}_k^n = \mathfrak{N}_k(\Gamma^n)$ of cusp forms. Now let $\mathbf{V} \subset \mathfrak{M}_k^n$ be an arbitrary invariant subspace. We set $\mathbf{V}_1 = \mathbf{V} \cap \mathfrak{N}_k^n$ and denote by $\mathbf{V}_2 = \mathbf{V}_1^{\perp} = \{f \in \mathbf{V} | (F, G) = 0 \text{ for all } G \in \mathbf{V}_1\}$ the orthogonal complement of $\mathbf{V}_1$ in $\mathbf{V}$ with respect to the Petersson scalar product (1.70). It follows in a standard way from the properties of scalar product listed in Theorem 1.38 that the space $\mathbf{V}$ is the orthogonal direct sum of the subspaces $\mathbf{V}_1$ and $\mathbf{V}_2$:

$$\mathbf{V} = \mathbf{V}_1 \oplus \mathbf{V}_2. \tag{4.36}$$

By Proposition 4.4 and our assumption, the subspace $\mathbf{V}_1$ is the intersection of two invariant subspaces and so is also invariant with respect to all of the Hecke operators. By Lemma 4.6 with $q = 1$, each of the Hecke operators $|_k T$ with $T \in \mathcal{L}^n$ satisfies

$$(F|_k T, F') = (F, F'|_k T) \tag{4.37}$$

if one of the forms $F, F' \in \mathfrak{M}_k^n$ is a cusp form. It follows that the subspace $\mathbf{V}_2$ is again invariant under all of the Hecke operators, and it suffices to prove that this subspace has a basis of common eigenfunctions for all of the operators. Since the subspace $\mathbf{V}_2$ does not contain nonzero cusp forms, it follows from Lemma 4.21 that the Siegel operator Φ maps this subspace isomorphically onto its image $\mathbf{V}' = \mathbf{V}_2|\Phi$ in $\mathfrak{M}_k^{n-1}$. Since the mapping (4.33) with $\chi = 1$ is epimorphic for every prime p, it follows from Theorem 4.19 that the subspace $\mathbf{V}' \subset \mathfrak{M}_k^{n-1}$ is invariant with respect to all of the Hecke operators $|_k T'$ with $T' \in \mathcal{L}_p^{n-1}$ for all prime p and hence, by Theorem 3.37, with all $T' \in \mathcal{L}^{n-1}$. Assuming that the theorem has already been proved for invariant subspaces of $\mathfrak{M}_k^{n-1}$, we see that there is a basis $F_1', \dots, F_d'$ of the space $\mathbf{V}'$ consisting of common eigenfunctions. The inverse images $F_1 = F_1'|\Phi^{-1}, \dots, F_d = F_d'|\Phi^{-1}$ of the functions $F_1', \dots, F_d'$ in $\mathbf{V}_2$ clearly form a basis of $\mathbf{V}_2$, and each of the functions is a common eigenfunctions for all of the Hecke operators on $\mathfrak{M}_k^n$. In order to see the latter, it is sufficient to consider only the operators corresponding to elements of a local Hecke–Shimura ring $\mathcal{L}_p^n$. For $T \in \mathcal{L}_p^n$, by (4.31), we obtain

$$(F_i|_k T)|\Phi = F_i'|\Psi(\mathbf{T}) = \lambda(\Psi(\mathbf{T}))F_i' = (\lambda(\Psi(\mathbf{T}))F_i)|\Phi,$$

where $|_k = |_{k,1}$, $\mathbf{T}$ is the image of T in $\mathbf{L}_p^n$, and $\lambda(\Psi(\mathbf{T}))$ is a scalar. Hence $\big(F_i|_k T - \lambda(\Psi(\mathbf{T}))F_i\big)|\Phi = 0$, and so $F_i|_k T - \lambda(\Psi(\mathbf{T}))F_i = 0$. In order to complete the induction, it remains to prove the theorem for $n = 1$. If we again present an invariant subspace $\mathbf{V}$ of $\mathfrak{M}_k^1$ in the form (4.35) of the direct sum of two invariant subspaces, then since $\mathfrak{M}_k^0 = \mathbb{C}$, the dimension of $\mathbf{V}_2$ is 0 or 1. If $\dim \mathbf{V}_2 = 0$, then $\mathbf{V} \subset \mathfrak{N}_k^1$ and the assertion has already been proved. But if $\dim \mathbf{V}_2 = 1$, then every function of $\mathbf{V}_2$ is a common eigenfunction. □

Exercise 4.23. Show that all eigenvalues of all Hecke operators on $\mathfrak{M}_k^n$ are real numbers.

Exercise 4.24. Let p be a prime number and $F \in \mathfrak{M}_k^n$ a nonzero common eigenfunction of all Hecke operators corresponding to elements T of $\mathcal{L}_p^n$ with the eigenvalues $\lambda(T) = \lambda_F(T)$. Let

$$Q_{p,F}(v) = \sum_{0 \le i \le 2^n} (-1)^n \lambda(q_i^n(p)) v^i$$

be the spinor p-polynomial of F resulting from the replacement of coefficients of the spinor p-polynomial (3.112) with eigenvalues of corresponding Hecke operators acting on the eigenfunction F, and let $\alpha_0(p), \alpha_1(p), \ldots, \alpha_n(p)$ be parameters of the linear extension on $\widetilde{\mathcal{L}}_p^n$ of the homomorphism $T \mapsto \lambda(T)$ defined in Proposition 3.55. Prove the following formulas:

(1) $Q_{p,F}(v) = (1 - \alpha_0(p)v) \prod_{r=1}^n \prod_{1 \le i_1 < \cdots < i_r \le n} (1 - \alpha_0(p)\alpha_{i_1}(p) \cdots \alpha_{i_r}(p))$.

(2) $\alpha_0(p)^2 \alpha_1(p) \cdots \alpha_n(p) = p^{nk - \frac{n(n+1)}{2}}$.

(3) $p^{2^{n-1}\left(nk - \frac{n(n+1)}{2}\right)} v^{2^n} Q_{p,F}\left(1/p^{nk - \frac{n(n+1)}{2}} v\right) = Q_{p,\,F}(v)$.

(4) If F is not a cusp form, prove that the image $F|\Phi \in \mathfrak{M}_k^{n-1}$ of F under the Siegel operator is a common eigenfunction for all the Hecke operators corresponding to elements of $\mathcal{L}_p^{n-1}$ and that the following Zharkovskaya formulas hold:

$$Q_{p,\,F}(v) = Q_{p,\,F|\Phi}(v) Q_{p,\,F|\Phi}(p^{k-n} v).$$

[Hints: For part 2 first find $\lambda_F(\langle p \rangle_n)$ and then use (3.113); for part 4, use (4.32).]

Chapter 5
Euler Factorization of Radial Series

Representations of symplectic Hecke–Shimura rings on spaces of modular forms given by Hecke operators provide a tool to approach the multiplicative properties of modular forms, which presumable must reflect the multiplicative properties of the rings. It is not obvious in what terms the multiplicative properties of modular forms can be formulated, since the ordinary multiplication of the form leads out of the initial spaces and apparently has no relation to Hecke operators. One can ask instead, say, about multiplicative properties of Fourier coefficients of modular forms that are sets of complex numbers indexed by appropriate arithmetic sequences. For example, in the case of modular forms for the groups $\Gamma_0^n(q)$, Fourier coefficients are indexed by nonnegative rational integers if $n = 1$ and by classes of positive semidefinite integral quadratic forms in n variables with respect to proper integral equivalence if $n > 1$. Although the notion of multiplicative function of integral argument is quite common in number theory, there is no natural definition of such functions on classes of integral quadratic forms in $n > 1$ variables, since generally, there is no multiplication of the classes, except for some isolated cases such as Gauss composition of binary forms. A possible way out of the situation is to consider suitable generating series for values of the functions instead of the values themselves. Traditionally, starting from Euler, multiplicativity of arithmetic sequences is customarily expressed in the form of an Euler product factorization of the generating Dirichlet series. It turns out that in the situation of modular forms, suitable Dirichlet series constructed by Fourier coefficients of eigenfunctions of Hecke operators can be expressed through Dirichlet series formed by the corresponding eigenvalues. The latter, according to the multiplicative properties of the relevant Hecke–Shimura rings, may have an Euler product factorization. Such relations between Dirichlet series constructed by Fourier coefficients of the eigenfunctions and the Euler product formed with the help of the corresponding eigenvalues prove to be useful in both directions: on the one hand, they establish the multiplicativity of the Fourier coefficients, and on the other hand, the relations allow one to express the Euler products, called *zeta functions of modular forms,* in terms of underlying modular forms and so to relate their analytic properties with those of the forms, as has been noted in the introduction.

A. Andrianov, *Introduction to Siegel Modular Forms and Dirichlet Series*, Universitext, DOI 10.1007/978-0-387-78753-4_5,

5.1 Radial Series of Genus One and Zeta Functions

In this section,

$$F = F(z) \in \mathfrak{M}_k(q,\chi) = \mathfrak{M}_k(\Gamma_0(q),\chi) \tag{5.1}$$

is a modular form in one complex variable z belonging to the upper half-plane $\mathbb{H} = \mathbb{H}^1 = \{z = x+iy \in \mathbb{C} | y > 0\}$ of a positive integral weight k and a character χ of the form (2.3) for the group $\Gamma_0(q) = \Gamma_0^1(q)$, where $q \in \mathbb{N}$ and χ is a Dirichlet character modulo q. We shall write the Fourier expansion (1.35) of F in the form

$$F(z) = \sum_{2a\in\mathbb{E}^1,\ a\geq 0} f(2a)e^{\pi i 2az} = \sum_{a=0}^{\infty} \mathrm{f}(a)e^{2\pi iaz} \tag{5.2}$$

with Fourier coefficients $\mathrm{f}(a) = f(2a)$ indexed by all nonnegative integers.

Fourier Coefficients of Eigenfunctions and Eigenvalues. It turns out that for modular forms in one variable, there are close relations between individual Fourier coefficients of the eigenfunctions and the corresponding eigenvalues.

Proposition 5.1. *Let F be a modular form* (5.1) *with Fourier expansion* (5.2). *Suppose that F is an eigenfunction for the Hecke operator* $|T(m) = |_{k,\chi}T(m)$, *where* $T(m) \in \mathcal{L}_0(q) = \mathcal{L}_0^1(q)$ *with m prime to q is an element of the form* (3.82) *for* $K = \Gamma_0(q)$*:*

$$F|T(m) = \lambda(m)F. \tag{5.3}$$

Then the Fourier coefficients $\mathrm{f}(a)$ *of F and the eigenvalue* $\lambda(m) = \lambda_F(m)$ *are linked by the relations*

$$\lambda(m)\mathrm{f}(a) = \sum_{d|m,\ a} \chi(d)d^{k-1}\mathrm{f}\left(\frac{ma}{d^2}\right) \quad \textit{for all } a \geq 0. \tag{5.4}$$

Proof. By Lemma 4.12, Proposition 3.61, and Lemma 4.14, the Fourier coefficients $(f \mid T(m))(a)$ of the function $F|T(m)$ can be written in the form

$$\begin{aligned}
(f|T(m))(a) = (f|\mathbf{T}(m))(a) &= \left(f|\sum_{d,d_1\in\mathbb{N},dd_1=m} \Pi_+^1(d)\Pi_-^1(d_1)\right)(a) \\
&= \sum_{d,\ d_1\in\mathbb{N},dd_1=m} ((f|\Pi_+^1(d))|\Pi_-^1(d_1))(a) \\
&= \sum_{dd_1=m,d|a} \chi(d_1)d_1^{k-1}(f|\Pi_+^1(d))(a/d_1) \\
&= \sum_{dd_1=m,d|a} \chi(d_1)d_1^{k-1}\mathrm{f}(da/d_1) = \sum_{d|m,\ a} \chi(d)d^{k-1}\mathrm{f}(ma/d^2).
\end{aligned}$$

On the other hand, by (5.3), this Fourier coefficient is equal to $\lambda(m)\mathrm{f}(a)$. The relation (5.4) follows. □

First of all, the relations (5.4) are useful for studying properties of the eigenvalues.

Theorem 5.2. *The eigenvalues $\lambda(m) = \lambda_F(m)$ of the Hecke operators $|T(m) = |_{k,\chi}T(m)$ corresponding to a nonzero common eigenfunction $F \in \mathfrak{M}_k(q,\chi)$ of the operators with* $\gcd(m,q) = 1$ *have the following properties:*

(1) *For every m and m' prime to q, the eigenvalues satisfy the multiplicative relations*

$$\lambda(m)\lambda(m') = \sum_{d|m,\ m'} \chi(d)d^{k-1}\lambda\left(\frac{mm'}{d^2}\right). \tag{5.5}$$

(2) *If F is a cusp form, then the eigenvalues satisfy the inequalities*

$$|\lambda(m)| \le cm^{k/2}, \tag{5.6}$$

where the constants $c = c_F$ depend only on F.

Proof. Let $f(0), f(1), \dots$ be the Fourier coefficients of F. Since $k > 0$, the form F is not a constant, and there are positive integers a with $f(a) \neq 0$. Let $\kappa = \kappa(F)$ be the smallest of such integers, and let δ be the greatest divisor of κ that is coprime to q. Since the numbers δ and κ/δ are coprime, it follows from (5.4) with $m = \delta$ and $a = \kappa/\delta$ that $f(\kappa) = f(\delta \cdot \kappa/\delta) = \chi(\delta)\lambda(\delta)f(\kappa/\delta)$. Thus $f(\kappa/\delta) \neq 0$, and so $\delta = 1$, which means that

$$\kappa(F)|q^\infty, \text{ in particular, } \kappa(F) = 1 \text{ if } q = 1. \tag{5.7}$$

Hence by (5.4) with $a = \kappa$, we obtain the relations

$$\lambda(m) = f(m\kappa)/f(\kappa) \text{ for all } m \text{ prime to } q. \tag{5.8}$$

By (5.8) and (5.4), we get

$$\begin{aligned}\lambda(m)\lambda(m') &= \lambda(m)f(m'\kappa)/f(\kappa)\\ &= \sum_{d|m,\ m'\kappa} \chi(d)d^{k-1}f\left(\frac{mm'\cdot\kappa}{d^2}\right)/f(\kappa) = \sum_{d|m,\ m'} \chi(d)d^{k-1}\lambda\left(\frac{mm'}{d^2}\right),\end{aligned}$$

since common divisors of m and $m'\kappa$ must divide m'.

The estimates (5.6) follow from (5.8) and (1.46). □

It follows from the estimate (1.52) of the Fourier coefficients and relations (5.8) that, for, the eigenvalues of an arbitrary eigenfunction satisfy the inequalities

$$|\lambda(m)| \le c_F m^k \qquad (m \in \mathbb{N}, \gcd(m,q) = 1). \tag{5.9}$$

Exercise 5.3. Prove the relations (5.5), using the relations of Exercise 3.47 and relations (4.26).

Exercise 5.4. In the notation of Theorem 5.2, show that if $f(0) \neq 0$, then

$$\lambda(m) = \sum_{d|m} \chi(d) d^{k-1} \qquad (m \in \mathbb{N}, \gcd(m,q) = 1).$$

Exercise 5.5. In the notation of Theorem 5.2, prove that the eigenvalues $\lambda(m) = \lambda_F(m)$ corresponding to the Eisenstein series $F = E_k(z) \in \mathfrak{M}_k$ (see Exercise 4.9) are given by the formula

$$\lambda(m) = \sum_{d|m} d^{k-1} \qquad (m \in \mathbb{N});$$

use the formula to show that the Fourier coefficients $f(a)$ of E_k satisfy the relations

$$f(a) = f(1) \sum_{d|a} d^{k-1} \qquad (a \in \mathbb{N})$$

(compare Exercise 1.24).

Dirichlet Series and Zeta Functions of Modular Forms. Here we reformulate the relations (5.4) and (5.5) in the terms of generating Dirichlet series. Let us look first at a formal consequence of relations (5.5) in terms of Dirichlet series.

Lemma 5.6. *Let $q \in \mathbb{N}$ then the relations* (5.5) *for a nonzero function*

$$\lambda : \{m \in \mathbb{N} | \gcd(m,q) = 1\} \mapsto \mathbb{C}$$

are equivalent to the following formal Euler factorization of the Dirichlet series with coefficients $\lambda(m)$:

$$\sum_{m \in \mathbb{N}, \gcd(m,q)=1} \frac{\lambda(m)}{m^s} = \prod_{p \in \mathbb{P}, p \nmid q} \left(1 - \frac{\lambda(p)}{p^s} + \frac{\chi(p)p^{k-1}}{p^{2s}}\right)^{-1}. \tag{5.10}$$

Proof. The relations (5.5) are equivalent to the relations

$$\lambda(m)\lambda(m') = \lambda(mm') \quad \text{if } \gcd(mm',q) = 1 \text{ and } \gcd(m,m') = 1 \tag{5.11}$$

together with the relations

$$\lambda(p)\lambda(p^\delta) = \lambda(p^{\delta+1}) + \chi(p)p^{k-1}\lambda(p^{\delta-1}) \quad \text{for each } p \in \mathbb{P},\ p \nmid q \text{ and } \delta \geq 1. \tag{5.12}$$

The relations (5.11) are obviously equivalent to the formal factorization

$$\sum_{m \in \mathbb{N}, \gcd(m,\, q)=1} \frac{\lambda(m)}{m^s} = \prod_{p \in \mathbb{P}, p \nmid q} \left(\sum_{\delta=0}^{\infty} \frac{\lambda(p^\delta)}{p^{\delta s}}\right),$$

whereas the recursion relations (5.12) can be written as the identities for formal power series of the form

$$(1-\lambda(p)t+\chi(p)p^{k-1}t^2)\left(\sum_{\delta=0}^{\infty}\lambda(p^\delta)t^\delta\right)=1$$

or, by substituting $t=p^{-s}$, as the summation formulas

$$\sum_{\delta=0}^{\infty}\frac{\lambda(p^\delta)}{p^{\delta s}}=\left(1-\frac{\lambda(p)}{p^s}+\frac{\chi(p)p^{k-1}}{p^{2s}}\right)^{-1}. \quad \square$$

Exercise 5.7. Under the assumptions of Lemma 5.6 prove the following formal Euler factorization of the formal Rankin Dirichlet series:

$$\sum_{m\in\mathbb{N},\,\gcd(m,q)=1}\frac{\lambda(m^2)}{m^s}$$
$$=\prod_{p\in\mathbb{P},\,p\nmid q}\left(1+\frac{\chi(p)p^{k-1}}{p^s}\right)\left(1-\frac{\lambda(p^2)-\chi(p)p^{k-1}}{p^s}+\frac{\chi(p^2)p^{2k-2}}{p^{2s}}\right)^{-1}.$$

[Hint: First sum the formal power series $\sum_{\delta\geq 0}\lambda(p^{2\delta})t^\delta$.]

Let $F\in\mathfrak{M}_k(q,\chi)$ be a nonzero common eigenfunction for all regular Hecke operators, i.e., operators $|T=|_{k,\chi}T$ with $T\in\mathcal{L}_0(q)$ and in particular, $F|T(m)=\lambda(m)F$ for all m prime to q. Then we call the Dirichlet series

$$\zeta_r(s,F)=\sum_{m\in\mathbb{N},\,\gcd(m,q)=1}\frac{\lambda(m)}{m^s} \tag{5.13}$$

the *regular zeta function of the eigenfunction* F. If $q=1$, we omit the adjective "regular" and the lower index r and speak just about the *zeta function* $\zeta(s,F)$ *of* F. From Lemma 5.6 and the estimate (5.6), by standard calculus, we have the following result.

Proposition 5.8. *Let a nonzero cusp form $F\in\mathfrak{N}_k(q,\chi)$ be a common eigenfunction for all the regular Hecke operators. Then the regular zeta function $\zeta_r(s,F)$ of F converges absolutely and uniformly in each right half-plane $\Re s\geq\frac{k}{2}+1+\varepsilon$ with $\varepsilon>0$ and has the factorization into an absolutely and uniformly convergent Euler product of the form*

$$\zeta_r(s,F)=\prod_{p\in\mathbb{P},\,p\nmid q}\left(1-\frac{\lambda(p)}{p^s}+\frac{\chi(p)p^{k-1}}{p^{2s}}\right)^{-1}. \tag{5.14}$$

For an arbitrary nonzero eigenform $F\in\mathfrak{M}_k(q,\chi)$, by Lemma 5.6 and (5.9), the factorization (5.14) holds in every half-plane $\Re s\geq k+1+\varepsilon$ with $\varepsilon>0$.

Exercise 5.9. Show that the zeta function of the Eisenstein series $E_k(z)$ has the form

$$\zeta(s,E_k)=\zeta(s)\zeta(s-k+1),$$

where $\zeta(s)$ is the Riemann zeta function.

In Section 2.1 we associated with each cusp form of the space $\mathfrak{N}_k(q,\chi) = \mathfrak{N}_k^1(q,\chi)$ the radial Dirichlet series of the form (2.7) constructed by means of Fourier coefficients of F, which, according to Theorem 2.1, have nice analytic properties. On the other hand, by Theorem 4.7, the space $\mathfrak{N}_k(q,\chi)$ is spanned by eigenfunctions for all of the regular Hecke operators, and according to Proposition 5.8, the regular zeta function of every such eigenfunction is a Dirichlet series with Euler product factorization. A natural question arises whether radial Dirichlet series and Dirichlet series with Euler product associated with an eigenform are related to each other, or in other words, whether the Dirichlet series constructed by Fourier coefficients of an eigenform have an Euler product factorization. The following theorem gives a partly positive answer.

Theorem 5.10. *Let a nonzero cusp form $F \in \mathfrak{N}_k(q,\chi)$ with Fourier coefficients $f(a)$ be a common eigenfunction for all regular Hecke operators, in particular, $F|_{k,\chi}T(m) = \lambda(m)F$ for every m prime to q. Then the identity*

$$\sum_{a=1}^{\infty} \frac{f(a)}{a^s} = \left(\sum_{a|q^\infty} \frac{f(a)}{a^s}\right) \prod_{p\in\mathbb{P},\, p\nmid q} \left(1 - \frac{\lambda(p)}{p^s} + \frac{\chi(p)p^{k-1}}{p^{2s}}\right)^{-1} \tag{5.15}$$

is valid in the half-plane $\Re s > \frac{k}{2}+1$, where the series and product are absolutely convergent. In particular, if $q=1$, then

$$\sum_{a=1}^{\infty} \frac{f(a)}{a^s} = f(1)\prod_{p\in\mathbb{P}} \left(1 - \frac{\lambda(p)}{p^s} + \frac{\chi(p)p^{k-1}}{p^{2s}}\right)^{-1} \qquad (\Re s > \frac{k}{2}+1). \tag{5.16}$$

Proof. By Proposition 5.1, and we have

$$\sum_{a=1}^{\infty} \frac{f(a)}{a^s} = \sum_{a|q^\infty,\ \gcd(m,q)=1} \frac{f(am)}{(am)^s} = \left(\sum_{a|q^\infty} \frac{f(a)}{a^s}\right) \sum_{m,\, \gcd(m,q)=1} \frac{\lambda(m)}{m^s},$$

whence, by Proposition 5.8, we obtain the identity (5.15). □

For an arbitrary eigenform of all regular Hecke operators on the space $\mathfrak{M}_k(q,\chi)$, for the same reasons and by the corresponding estimates, the identity (5.15) holds in the half-plane $\Re s > k+1$.

We shall use the notation

$$Z_q(s,F) = \sum_{a|q^\infty} \frac{f(a)}{a^s} \tag{5.17}$$

for the q-part of the Dirichlet series $Z(s,F)$ of a modular form $F \in \mathfrak{M}_k(q,\chi)$ whose coefficients are the Fourier coefficients of F indexed only by positive integers dividing a power of q. In this and the above notation, the identity (5.15) takes the form

$$Z(s,F) = Z_q(s,F)\zeta_r(s,F) \quad \text{(resp., } Z(s,F) = f(1)\zeta(s,F) \text{ if } q=1\text{)}. \tag{5.18}$$

Exercise 5.11. Let $F_1, \dots, F_h$ be a basis of an invariant subspace of the space $\mathfrak{M}_k(q, \chi)$ under all regular Hecke operators, in particular,

$$F_i|_{k,\chi}T(m) = \sum_{1 \le j \le h} \lambda_{ij}(m)F_j \quad (i = 1, \dots, h; \gcd(m, q) = 1).$$

Let $\mathbf{f}(a)$ with $a = 0, 1, \dots$ be the h-column with i-entry for $i = 1, \dots, h$ equal to the Fourier coefficient $f_i(a)$ of F_i, and let $\Lambda(m) = (\lambda_{ij}(m))$. Prove the following relations:

$$\Lambda(m)\mathbf{f}(a) = \sum_{d|m,a} \chi(d)d^{k-1}\mathbf{f}\left(\frac{ma}{d^2}\right) \quad (a = 0, 1, \dots; \gcd(m, q) = 1);$$

$$\Lambda(m)\Lambda(m') = \sum_{d|m,m'} \chi(d)d^{k-1}\Lambda\left(\frac{mm'}{d^2}\right) \quad (\gcd(mm', q) = 1);$$

$$\sum_{a=1}^{\infty} a^{-s}\mathbf{f}(a) = \prod_{p \in \mathbb{P}, p \nmid q} \left(1_h - p^{-s}\Lambda(p) + \chi(p)p^{k-1-2s}1_h\right)^{-1} \left(\sum_{a|q^\infty} \frac{\mathbf{f}(a)}{a^s}\right).$$

Exercise 5.12. Let $F \in \mathfrak{M}_k$ be a nonzero eigenform with $F|_kT(m) = \lambda(m)F$ for all $m \in \mathbb{N}$. Prove that if the eigenvalues satisfy the Ramanujan–Petersson inequality $|\lambda(p)| \le 2p^{\frac{k-1}{2}}$ for each prime number p, then for arbitrary $m \in \mathbb{N}$, the eigenvalues satisfy

$$|\lambda(m)| \le \tau(m)m^{\frac{k-1}{2}},$$

where $\tau(m)$ is the number of positive divisors of m.

[Hint: use Exercise 4.23.]

Functional Equation for the Zeta Function. According to Theorem 2.1, the radial Dirichlet series $Z(s, F)$ of a cusp form $F \in \mathfrak{M}_k(q, \chi)$ has an analytic continuation over the whole s-plane such that the function $\Phi(s, F) = (2\pi)^{-s}\Gamma(s)Z(s, F)$, where $\Gamma(s)$ is the gamma function, is holomorphic everywhere and satisfies the functional equation

$$\Phi(k - s, F) = i^k q^{s-k+1}\Phi(s, F^*), \quad \text{with } F^* = F|_k \begin{pmatrix} 0 & -1 \\ 0 & q \end{pmatrix} \in \mathfrak{M}_k(q, \overline{\chi}). \tag{5.19}$$

The following proposition complements properties of the Dirichlet series entering into the functional equation.

Proposition 5.13. *If a modular form $F \in \mathfrak{M}_k(q, \chi)$ is an eigenfunction for all the Hecke operators $|_{k,\chi}T(m)$ with m prime to q, and $\lambda(m)$ are the corresponding eigenvalues, then the form $F^* \in \mathfrak{M}_k(q, \overline{\chi})$ is an eigenfunction for all Hecke operators $|_{k,\overline{\chi}}T(m)$ with m prime to q, and the corresponding eigenvalues are equal to $\overline{\chi}(m)\lambda(m)$; the regular zeta function of the eigenfunction F^* has properties similar* *to those of the regular zeta function of F listed in Theorem* 5.10; *in particular, it has an Euler factorization of the form*

$$\begin{aligned}\zeta_r(s,F^*) &= \sum_{m\in\mathbb{N},\,\gcd(m,q)=1} \frac{\overline{\chi}(m)\lambda(m)}{m^s} \\ &= \prod_{p\in\mathbb{P},\,p\nmid q}\left(1-\frac{\overline{\chi}(p)\lambda(p)}{p^s}+\frac{\overline{\chi}(p)p^{k-1}}{p^{2s}}\right)^{-1}\end{aligned} \tag{5.20}$$

Proof. To prove the first part it is sufficient to show that

$$F|_k\begin{pmatrix}0&-1\\0&q\end{pmatrix}|_{k,\overline{\chi}}T(p)=\overline{\chi}(p)F|_{k,\chi}T(p)|_k\begin{pmatrix}0&-1\\0&q\end{pmatrix} \tag{5.21}$$

for every $F\in\mathfrak{M}_k(q,\chi)$ and prime number p not dividing q. By Lemma 3.49, the matrix *smap*, 0, 0, 1. and matrices $\left(\begin{smallmatrix}1&b\\0&p\end{smallmatrix}\right)$ with $b=0,1,\ldots,p-1$ form a system of representatives of left cosets modulo $K=\Gamma_0^1(q)$ contained in the double coset $T(p)$. If $p\nmid b$, then there is $b'\in\{1,\ldots,p-1\}$ such that $1+qbb'\equiv 0 \pmod p$. Hence for $b=1,2,\ldots,p-1$, we get the relation

$$\begin{pmatrix}0&-1\\q&0\end{pmatrix}\begin{pmatrix}1&b\\0&p\end{pmatrix}=\begin{pmatrix}p&-b'\\-qb&(1+qbb')/p\end{pmatrix}\begin{pmatrix}1&b'\\0&p\end{pmatrix}\begin{pmatrix}0&-1\\q&0\end{pmatrix},$$

where the first factor on the right belongs to the group K. Hence by (4.13), (4.14), and (1.31), we obtain

$$\begin{aligned}F|_k\begin{pmatrix}0&-1\\0&q\end{pmatrix}|_{k,\overline{\chi}}T(p) &= \overline{\chi}(p)F|_k\begin{pmatrix}0&-1\\0&q\end{pmatrix}\begin{pmatrix}p&0\\0&1\end{pmatrix} \\ &\quad +F|_k\begin{pmatrix}0&-1\\0&q\end{pmatrix}\begin{pmatrix}1&0\\0&p\end{pmatrix}+\sum_{b=1}^{p-1}F|_k\begin{pmatrix}0&-1\\0&q\end{pmatrix}\begin{pmatrix}1&b\\0&p\end{pmatrix} \\ &= \overline{\chi}(p)F|_k\begin{pmatrix}1&0\\0&p\end{pmatrix}\begin{pmatrix}0&-1\\q&0\end{pmatrix}+F|_k\begin{pmatrix}p&0\\0&1\end{pmatrix}\begin{pmatrix}0&-1\\q&0\end{pmatrix} \\ &\quad +\sum_{b=1}^{p-1}F|_k\begin{pmatrix}p&-b'\\-qb&(1+qbb')/p\end{pmatrix}\begin{pmatrix}1&b'\\0&p\end{pmatrix}\begin{pmatrix}0&-1\\q&0\end{pmatrix} \\ &= \overline{\chi}(p)\Bigg(F|_{k,\chi}\begin{pmatrix}1&0\\0&p\end{pmatrix}+F|_{k,\chi}\begin{pmatrix}p&0\\0&1\end{pmatrix} \\ &\quad +\sum_{b=1}^{p-1}F|_{k,\chi}\begin{pmatrix}1&b'\\0&p\end{pmatrix}\Bigg)|_k\begin{pmatrix}0&-1\\q&0\end{pmatrix},\end{aligned}$$

which proves the relation (5.21). The Euler factorization (5.20) follows from Proposition 5.8 with F^* in place of F. □

It follows from the proposition and Theorem 5.10 that the function $\Phi(s,F)$ and $\Phi(s,F^*)$ in the functional equation (5.19), for a nonzero eigenfunction of all regular

Hecke operators on the space $\mathfrak{N}_k(q,\chi)$, in the half-plane $\Re s > \frac{k}{2}+1$, can be written in the form

$$\Phi(s,F) = (2\pi)^{-s}\Gamma(s)Z_q(s,F)\zeta_r(s,F)$$

and

$$\Phi(s,F^*) = (2\pi)^{-s}\Gamma(s)Z_q(s,F^*)\zeta_r(s,F^*),$$

where $\zeta_r(s,F)$ and $\zeta_r(s,F^*)$ are the regular zeta functions of the eigenforms F and F^*, and $Z_q(s,F), Z_q(s,F^*)$ are the corresponding series (5.17). Thus, the functional equation can be considered as a functional equation for the zeta functions. In particular, since for $q=1$ clearly $F^*=F$ and $Z_1(s,F)=Z_1(s,F^*)=f(1)\neq 0$, we get the following theorem.

Theorem 5.14. *The zeta function $\zeta(s,F)$ of a nonzero eigenform $F\in\mathfrak{N}_k$ has a meromorphic continuation to the whole s-plane; the function*

$$\Phi(s,F) = (2\pi)^{-s}\Gamma(s)\zeta(s,F)$$

is holomorphic and satisfies the functional equation

$$\Phi(k-s,F) = i^k\Phi(s,F). \tag{5.22}$$

Exercise 5.15. Show that the zeta function $\zeta(s,E_k)$ of the Eisenstein series E_k with even k has a meromorphic continuation over the whole s-plane, and that the function $\Phi(s,E_k)=(2\pi)^{-s}\Gamma(s)\zeta(s,E_k)$ is analytic except for two simples poles at $s=0$ and $s=k$ and satisfies the functional equation $\Phi(k-s,F)=i^k\Phi(s,F)$.

[Hint: Use Exercise 2.2, or Exercise 5.9 and properties of the Riemann zeta function.]

In framework of the theory of *old and new forms* initiated by A.O.L. Atkin and J. Lehner, which we do not touch on in this book, it turns out that under certain conditions, the q-partial Dirichlet series $Z_q(s,F)$ is a finite Euler product extending only over prime divisors of q.

5.2 Binary Radial Series and Gaussian Composition

In this section we fix a nonzero modular form

$$F = F(Z) \in \mathfrak{M}_k^2(q,\chi) = \mathfrak{M}_k(\Gamma_0^2(q),\boldsymbol{\chi}) \tag{5.23}$$

of positive integral weight k and character χ modulo $q\in\mathbb{N}$ for the group $\Gamma_0^2(q)$ defined on the upper half-plane $\mathbb{H}^2$ of genus 2. The Fourier expansion (1.35) of such an F takes the form

$$F(Z) = \sum_{A\in\mathbb{E}^2,\, A\geq 0} f(A)e^{\pi i\sigma(AZ)} \tag{5.24}$$

with Fourier coefficients indexed by matrices A of integral nonnegative semidefinite binary quadratic forms

$$\mathbf{a}(x,y) = \frac{1}{2}{}^tXAX = ax^2 + bxy + cy^2 \Leftrightarrow A = \begin{pmatrix} 2a & b \\ b & 2c \end{pmatrix} \quad \left(X = \begin{pmatrix} x \\ y \end{pmatrix} \right). \tag{5.25}$$

Depending on circumstances, we shall use the language of even matrices or of integral quadratic forms with similarly defined basic notation. So we say that a number $d \in \mathbb{N}$ *divides* the form or matrix (5.25) if it divides the coefficients a, b, c, and the greatest divisor

$$e = e(\mathbf{a}) = e(A) = \gcd(a, b, c) \tag{5.26}$$

is *the divisor* of a nonzero form or matrix. If the divisor is equal to 1, the form and matrix are called *primitive*. The number $\det \mathbf{a} = \det A$ is the *determinant* of the form, and the number

$$\delta = \delta(\mathbf{a}) = \delta(A) = -\det a = b^2 - 4ac \tag{5.27}$$

is the *discriminant* of the form and the matrix. If A and A' are even matrices and $A' = {}^t\lambda A \lambda$ with $\lambda \in \Lambda_+^2 = \mathrm{SL}_2(\mathbb{Z})$, then matrices A and A' and the corresponding quadratic forms are called *(properly) equivalent*, and we write then

$$A' \sim A \quad \text{and} \quad \mathbf{a}' \sim \mathbf{a}.$$

Equivalent even matrices or integral forms make up a *(proper) class*. We denote by $\{A\}_+$ and $\{\mathbf{a}\}_+$ the proper classes of A and $\mathbf{a}$, respectively. All even matrices and integral forms from a fixed class have the same signature, divisor, and discriminant.

According to (4.16), the Fourier coefficients $f(A)$ of every modular form for $\Gamma_0^2(q)$ depend only on the proper class of the matrix A.

Fourier Coefficients of Eigenfunctions and Eigenvalues. As in the case of modular forms in one variable, Fourier coefficients of the eigenfunctions of genus 2 are closely related to the corresponding eigenvalues, but the relations are not so simple.

Proposition 5.16. *Let F be a modular form* (5.23) *with Fourier expansion* (5.24). *Suppose that F is an eigenfunction for the Hecke operator $|T(m) = |_{k,\chi}T(m)$, where $T(m) \in \mathcal{L}_0(q) = \mathcal{L}_0^2(q)$ with m prime to q is an element of the form* (3.82) *for $K = \Gamma_0^2(q)$:*

$$F|T(m) = \lambda(m)F. \tag{5.28}$$

Then the Fourier coefficients $f(A)$ of F and the eigenvalue $\lambda(m) = \lambda_F(m)$ are linked by the relations

$$\lambda(m)f(A) = \sum_{\substack{d,d_1,d_2 \in \mathbb{N};\, A \in d_2\mathbb{E}^2, \\ dd_1d_2 = m}} \chi(d_1)\chi(d_2^2)d_1^{k-2}d_2^{2k-3} \times \sum_{\substack{D \in \Lambda_+ \backslash \Lambda_+ \mathrm{diag}(1,d_1)\Lambda_+; \\ d_2^{-1}A[{}^tD] \in d_1\mathbb{E}^2}} f\left(\frac{d}{d_1d_2}A[{}^tD]\right), \tag{5.29}$$

where $\Lambda_+ = \Lambda_+^2$.

Proof. For the Fourier coefficient of the function $F|T(m)$ indexed by a matrix $A \in \mathbb{E}^2$, by formula (3.132) of Proposition 3.61, we obtain the formula

$$\lambda(m)f(A) = (f|T(m))(A) = \sum_{d,d_1,d_2\in\mathbb{N},\, dd_1d_2=m} (f|\Pi_+(d)\Pi(d_1)\Pi_-(d_2))(A),$$

where $\Pi_\pm = \Pi^2_\pm$. By formula (4.24) of Lemma 4.14, we get

$$(f|\Pi_+(d)\Pi(d_1)\Pi_-(d_2))(A) = \left((f|\Pi_+(d)|\Pi(d_1))|\Pi_-(d_2)\right)(A)$$
$$= \begin{cases} \chi(d_2)^2 d_2^{2k-3}(f|\Pi_+(d)|\Pi(d_1))(d_2^{-1}A) & \text{if } A \in d_2\mathbb{E}^2, \\ 0 & \text{if } A \notin d_2\mathbb{E}^2. \end{cases}$$

Further, by formulas (4.27) and (4.25) of Lemma 4.14, we conclude that

$$\left((f|\Pi_+(d))|\Pi(d_1)\right)(A')$$
$$= \chi(d_1)d_1^{k-2} \sum_{\substack{D\in\Lambda_+\backslash\Lambda_+ \operatorname{diag}(1,\, d_1)\Lambda_+;\\ A'[{}^tD]\in d_1\mathbb{E}^2}} (f|\Pi(d))\left(d_1^{-1}A'[{}^tD]\right)$$
$$= \chi(d_1)d_1^{k-2} \sum_{\substack{D\in\Lambda_+\backslash\Lambda_+ \operatorname{diag}(1,\, d_1)\Lambda_+;\\ A'[{}^tD]\in d_1\mathbb{E}^2}} f\left(\frac{d}{d_1}A'[{}^tD]\right).$$

These formulas imply the formulas (4.29). □

Although formulas (5.29) look rather complicated, we shall need only certain particular cases.

Corollary 5.17. *Let $F \in \mathfrak{M}^2_k(q,\chi)$ be a nonzero modular form with Fourier expansion (5.24) satisfying $F|T(m) = \lambda(m)F$ for all of the Hecke operators $|T(m) = |_{k,\chi}T(m)$ with m prime to q. Then for every primitive matrix $A \in \mathbb{E}^2$ and integer ν coprime to q, the Fourier coefficients of F and the eigenvalues $\lambda(m) = \lambda_F(m)$ are linked by the relation*

$$\lambda(m)f(\nu A) = \sum_{d,d_1\in\mathbb{N},\; dd_1=m} \chi(d_1)d_1^{k-2} \sum_{\substack{D\in\Lambda_+\backslash\Lambda_+ \operatorname{diag}(1,\, d_1)\Lambda_+,\\ A[{}^tD]\in d_1\mathbb{E}^2}} f\left(dd_1^{-1}\nu A[{}^tD]\right). \tag{5.30}$$

Proof. With νA in place of A, where A and ν satisfy the stated conditions, the formula (5.29) turns into (5.30), since for primitive A, the index d_2 must be equal to 1, and the conditions $\nu A[{}^tD] \in d_1\mathbb{E}^2$ and $A[{}^tD] \in d_1\mathbb{E}^2$ are obviously equivalent. □

Let us look at some attractive formal implementations of these last relations. Note that for every matrix D entering the inner sum on the right, the matrix $d_1^{-1}A[{}^tD]$ is even and has the same (positive) determinant as the matrix A. Let us suppose for a moment that the discriminant of a primitive nonsingular matrix A is a so-called single-class discriminant, i.e., such that each even primitive matrix A' of order 2 with the same discriminant is (properly) equivalent to A. Then, since Fourier coefficients depend only on the classes of indexing matrices, the last formula takes the shape

$$\lambda(m)f(\nu A) = \sum_{d,d_1\in\mathbb{N},\, dd_1=m} f(d\nu A)\chi(d_1)d_1^{k-2}\rho(d_1,A),$$

where

$$\rho(d,A) = \#\left\{D\in\Lambda_+\backslash\Lambda_+\mathrm{diag}(1,d)\Lambda_+ \;\middle|\; A[{}^tD]\in d\mathbb{E}^2\right\}. \tag{5.31}$$

Now if we divide both parts of the last relation by $m^s = d^s d_1^s$, considered just a formal quasicharacter of the semigroup $\mathbb{N}$, and then sum up these relations over all m prime to q, then as a result we get the following identity for formal Dirichlet series:

$$\begin{aligned} f(\nu A) &\sum_{m\in\mathbb{N},\,\gcd(m,\,q)=1} \frac{\lambda(m)}{m^s} \\ &= \sum_{d,d_1\in\mathbb{N},\,\gcd(dd_1,\,q)=1} \frac{f(d\nu A)}{d^s}\,\frac{\chi(d_1)d_1^{k-2}\rho(d_1,\,A)}{d_1^s} \\ &= \sum_{d\in\mathbb{N},\,\gcd(d,\,q)=1} \frac{f(d\nu A)}{d^s} \sum_{d_1\in\mathbb{N},\,\gcd(d_1,\,q)=1} \frac{\chi(d_1)\rho(d_1,A)}{d_1^{s-k+2}}. \end{aligned} \tag{5.32}$$

Therefore, by dividing both sides by ν^s and summing up over all $\nu\in\mathbb{N}$ with $\nu|q^\infty$, since $n=d\nu$ ranges over all of $\mathbb{N}$, we obtain for every primitive matrix A of single-class discriminant the relation

$$\begin{aligned} \sum_{\nu\in\mathbb{N},\,\nu|q^\infty} \frac{f(\nu A)}{\nu^s} &\sum_{m\in\mathbb{N},\,\gcd(m,\,q)=1} \frac{\lambda(m)}{m^s} \\ &= \sum_{n=1}^{\infty} \frac{f(nA)}{n^s} \sum_{d\in\mathbb{N},\,\gcd(d,\,q)=1} \frac{\chi(d)\rho(d,A)}{d^{s-k+2}}. \end{aligned} \tag{5.33}$$

It follows from Theorem 3.46 and Proposition 3.53 that the Dirichlet series on the left with coefficients $\lambda(m)$ has an Euler product factorization. Thus, an identity of the form (5.33), on the one hand, expresses this Euler product by means of a radial Dirichlet series on the right and so opens a way to derive analytic properties of the product from those of the radial series, and on the other hand, it reveals multiplicative properties of Fourier coefficients of the eigenfunction in the form of a (partial) Euler product factorization of radial series (we shall see later that the second Dirichlet series on the right also has an Euler product factorization). Note that in order to reveal multiplicativity of the Fourier coefficients, one needs similar identities for a variety of radial series, whereas in order to study analytic properties of the Euler product, it suffices to take such an identity only for a single nonzero radial series. Both of these problems will be considered below.

Exercise 5.18. In the notation and under assumptions of Proposition 5.16, if, moreover, $f(0_2)\neq 0$, prove that

$$\lambda(m) = \sum_{d,d_1,d_2\in\mathbb{N};\, dd_1d_2=m} \chi(d_1)\chi(d_2^2)d_1^{k-2}d_2^{2k-3}\#\{\Lambda_+\backslash\Lambda_+\operatorname{diag}(1,d_1)\Lambda_+\}.$$

***Π*-Operators and Multiplication of Quadratic Modules.** Along with the simple operations of multiplication and division of even matrices or corresponding quadratic forms by positive integers, the above formulas for the action of Hecke operators on Fourier coefficients of modular forms include more complicated operators $|\Pi(d)$ of the form (4.27). It turns out that these operators can also be interpreted with the help of a multiplication of quadratic forms, namely the multiplication in the sense of Gaussian composition.

We shall treat the Gaussian composition of quadratic forms using the equivalent language of multiplication of quadratic modules. First we have to recall some common definitions related to modules in a quadratic number field. For details see any textbook on algebraic numbers.

Let K be a quadratic extension of the field $\mathbb{Q}$ of rational numbers. An additive subgroup $\mathfrak{A}$ of K finitely generated over $\mathbb{Z}$ is called a *module* in K. Two modules $\mathfrak{A}$ and $\mathfrak{A}'$ are said to be *similar* or belong to one *class* $\{\mathfrak{A}'\} = \{\mathfrak{A}\}$ if $\mathfrak{A}' = \alpha\mathfrak{A}$ with a nonzero α of K; in this case one writes $\mathfrak{A}' \sim \mathfrak{A}$. The set of all modules in K is a disjoint union of the classes of similar modules. A module $\mathfrak{A}$ in K is said to be *full* if $\mathbb{Q}\mathfrak{A} = K$. Each full module in K has a $\mathbb{Z}-$basis of two elements. A full module in K that contains 1 and is a ring is called an *order* in K. Every order in K is contained in the maximal order $O = O(K)$, the ring of all (algebraic) integers of the field K over $\mathbb{Q}$. Let O' be an order and ω_1, ω_2 a basis of O'. Then the number $\delta(O') = \det(\operatorname{tr}(\omega_i\omega_j))$, where tr means the trace from K to $\mathbb{Q}$, is independent of the choice of the basis and called the *discriminant of* O'. The discriminant of the maximal order O is called the *discriminant* of the field K.

If $\mathfrak{A}$ is a module in K, then the ring

$$O(\mathfrak{A}) = \left\{\alpha \in K \,\middle|\, \alpha\mathfrak{A} \subset \mathfrak{A}\right\}$$

is called the *ring of multipliers of the module.* Similar modules have equal rings of multipliers. The ring of multipliers of a full module is an order in K. For each full module, there is a similar module contained in its ring of multipliers. For a full module $\mathfrak{A}$ with $O(\mathfrak{A}) = O'$, let α_1, α_2 and ω_1, ω_2 be bases of $\mathfrak{A}$ and O', respectively. Then the absolute value of the determinant of the transition matrix from the first basis to the second,

$$N(\mathfrak{A}) = |\det(a_{ij})|, \quad \text{where} \quad \alpha_i = \sum a_{ij}\omega_j,$$

is independent of the choice of bases, and is called the *norm* of $\mathfrak{A}$. If $\mathfrak{A} \subset O'$, then $N(\mathfrak{A}) = [O' : \mathfrak{A}]$ (the index of $\mathfrak{A}$ in O').

Let $\mathfrak{A}$ and $\mathfrak{A}'$ be two full modules in K. Then the set $\mathfrak{A}\mathfrak{A}'$ is again a full module in K, which is called the *product of* $\mathfrak{A}$ *and* $\mathfrak{A}'$. The norm of the product is equal to the product of norms:

$$N(\mathfrak{A}\mathfrak{A}') = N(\mathfrak{A})N(\mathfrak{A}').$$

For a full module $\mathfrak{A}$ in K we denote by $\overline{\mathfrak{A}}$ the module consisting of the conjugates $\bar{\alpha}$ over $\mathbb{Q}$ of the elements $\alpha \in \mathfrak{A}$. The *conjugate module* $\overline{\mathfrak{A}}$ is again a full module with the same ring of multipliers as that of $\mathfrak{A}$, and the following formula holds:

$$\mathfrak{A}\overline{\mathfrak{A}} = N(\mathbf{M})O(\mathfrak{A}).$$

All full modules in K with a fixed ring of multipliers O' form a commutative group under multiplication of modules. The quotient group of this group by the subgroup of modules similar to O' is a finite group $H(O')$ called the *class group* of O'. The order $h(O') = \#H(O')$ of the class group is the *class number of* O'.

Full modules $\mathfrak{A}$ in K contained in its ring of multipliers $O(\mathfrak{A}) = O'$ are called *regular ideals of* the ring O'. Each regular ideal of an order is uniquely (up to order) decomposable into a product of regular prime ideals.

Every quadratic extension of $\mathbb{Q}$ has the form $K = \mathbb{Q}(\sqrt{d_0})$, where $d_0 \neq 0, 1$ is a square-free integer. As a basis of the maximal order O of such K one can take the numbers 1 and ω, where

$$\omega = \begin{cases} (1+\sqrt{d_0})/2 & \text{if } d_0 \equiv 1 \pmod 4, \\ \sqrt{d_0} & \text{if } d_0 \equiv 2 \text{ or } 3 \pmod 4, \end{cases}$$

and the discriminant of the field is equal to

$$\mathbf{d} = \delta(K) = \delta(O) = \begin{cases} d_0 & \text{if } d_0 \equiv 1 \pmod 4, \\ 4d_0 & \text{if } d_0 \equiv 2 \text{ or } 3 \pmod 4. \end{cases}$$

In both cases one can write $K = \mathbb{Q}(\sqrt{\mathbf{d}})$. Any order O' of K has the form

$$O_l = \mathbb{Z} + l\omega\mathbb{Z},$$

where l is the index $[O : O']$. The discriminant of O_l is equal to $\mathbf{d}l^2$.

For $\alpha, \beta, \ldots \in K$ we shall denote by $\{\alpha, \beta, \ldots\}$ the full module in K generated over $\mathbb{Z}$ by $\alpha, \beta, \ldots$:

$$\{\alpha, \beta, \ldots\} = \mathbb{Z}\alpha + \mathbb{Z}\beta + \cdots.$$

The following lemma is often useful in computations with quadratic modules.

Lemma 5.19. *Let K be a quadratic field, $\gamma \in K$, $\gamma \notin \mathbb{Q}$, and let $a\gamma^2 + b\gamma + c = 0$, where a, b, and c are the rational integers satisfying* $\gcd(a,b,c) = 1$ *and $a > 0$. Then the module $\mathfrak{A} = \{1, \gamma\}$ satisfies the conditions*

$$O(\mathfrak{A}) = \{1, a\gamma\}, \quad N(\mathfrak{A}) = 1/a, \quad \delta(O(\mathfrak{A})) = b^2 - 4ac.$$

Proof. A number $\alpha = r + r'\gamma \in K$ with $r, r' \in \mathbb{Q}$ satisfies $\alpha\mathfrak{A} \subset \mathfrak{A}$ if and only if $\alpha = r + r'\gamma \in \mathfrak{A}$ and $\alpha\gamma = -r'c/a + (r - r'b/a)\gamma \in \mathfrak{A}$, which is equivalent to the inclusions $r, r', r'c/a, r'b/a \in \mathbb{Z}$. Since the numbers a, b, c are coprime, these inclusions are equivalent to the inclusions $r \in \mathbb{Z}$, $r' \in a\mathbb{Z}$, whence $O(\mathfrak{A}) = \{1, a\gamma\}$. For the basis $1, \gamma$ of $\mathfrak{A}$ and the basis $1, a\gamma$ of $O(\mathfrak{A})$, we have $N(\mathfrak{A}) = |\det(\mathrm{diag}(1, 1/a))| = 1/a$. The formula for the discriminant follows from the definition. □

The relation between quadratic modules and prime numbers is described by the following proposition.

Proposition 5.20. *Let $K = \mathbb{Q}(\sqrt{\mathbf{d}})$ be the quadratic field with discriminant* $\mathbf{d}$, *and O_l the order of K with discriminant $\mathbf{d}l^2$. Suppose that p is a rational prime number not dividing l. Then a full module $\mathfrak{A}$ in K satisfying the conditions*

$$O(\mathfrak{A}) = O_l, \quad \mathfrak{A} \subset O_l, \quad N(\mathfrak{A}) = p, \tag{5.34}$$

exists if and only if the congruence $x^2 \equiv \mathbf{d} \pmod{4p}$ is solvable. If the congruence is solvable and p does not divide $\mathbf{d}$, *then there are precisely two different modules $\mathfrak{P}$ and $\mathfrak{P}'$ satisfying* (5.34), *and $\mathfrak{P}' = \overline{\mathfrak{P}}$. If the congruence has a solution and p divides* $\mathbf{d}$, *then there is exactly one module $\mathfrak{P}$ satisfying* (5.34), *and $\overline{\mathfrak{P}} = \mathfrak{P}$. Finally, if the congruence has no solutions, then there is the unique complete module $\mathfrak{P}$ satisfying*

$$O(\mathfrak{P}) = O_l, \quad \mathfrak{P} \subset O_l, \quad N(\mathfrak{P}) = p^2, \tag{5.35}$$

and this module is $\mathfrak{P} = pO_l$.

All of the listed modules $\mathfrak{A} = \mathfrak{P}$ are regular prime ideals of the ring O_l.

Proof. Suppose that the module $\mathfrak{A}$ satisfies conditions (5.34). Then the index of $\mathfrak{A}$ in O_l is p, and so the least positive integer contained in $\mathfrak{A}$ equals p. Hence, the module $\mathfrak{A}$ has a basis of the form $p, p\gamma$, $\mathfrak{A} = \{p, p\gamma\}$, where $\gamma \in K$ and $\gamma \notin \mathbb{Q}$. The number γ satisfies the equation $a\gamma^2 + b\gamma + c = 0$, where a, b, and c are rational integers with $\gcd(a,b,c) = 1$ and $a > 0$. Then, by Lemma 5.19, we have $p = N(\mathfrak{A}) = N(pO_l)N(\{1,\gamma\}) = p^2/a$, whence $a = p$. On the other hand, $b^2 - 4ac = \delta(O(\mathfrak{A})) = \mathbf{d}l^2$, and so $b^2 \equiv \mathbf{d}l^2 \pmod{4p}$. Since $\mathbf{d} \equiv 0$ or $1 \pmod 4$ and $p \nmid l$, we conclude that the congruence $x^2 \equiv \mathbf{d} \pmod{4p}$ has a solution. Conversely, if this congruence has a solution, then there are rational integers b, c such that $b^2 - 4ac = \mathbf{d}$. It follows from the definition of $\mathbf{d}$ that $\gcd(p,b,c) = 1$, and so $\gcd(p, bl, cl^2) = 1$. Let γ be a root of the equation $px^2 + blx + cl^2 = 0$. Then the module $\mathfrak{A} = \{p, p\gamma\}$ satisfies (5.34).

Let $\mathfrak{A}_i = \{p, p\gamma_i\}$, where $i = 1,2$, be two modules satisfying (5.34). As has been shown, each of the numbers γ_i satisfies a relation $p\gamma_i^2 + b_i\gamma_i + c_i = 0$, where b_i, c_i are rational integers, $\gcd(p, b_i, c_i) = 1$, and $b_i^2 - 4pc_i = \mathbf{d}l^2$. In particular, $b_i^2 \equiv \mathbf{d}l^2 \pmod{4p}$, whence $b_1 \equiv b_2 \pmod{2p}$ or $b_1 \equiv -b_2 \pmod{2p}$. In the first case we obtain $b_2 = b_1 + 2tp$ and

$$\mathfrak{A}_2 = \{p, (-b_2 \pm l\sqrt{\mathbf{d}})/2\} = \{p, (-b_1 \pm l\sqrt{\mathbf{d}})/2 - tp\} = \{p, (-b_1 \pm l\sqrt{\mathbf{d}})/2\}.$$

The last module is equal to $\mathfrak{A}_1$ or $\overline{\mathfrak{A}}_1$. Similarly, in the second case, $b_2 = -b_1 + 2tp$ and

$$\mathfrak{A}_2 = \{p, (b_1 \pm l\sqrt{\mathbf{d}})/2 - tp\} = \{p, (-b_1 \mp l\sqrt{\mathbf{d}})/2\} = \mathfrak{A}_1 \text{ or } \overline{\mathfrak{A}}_1.$$

If $p|\mathbf{d}$, then $b_1 \equiv -b_1 \pmod{2p}$; hence $\overline{\mathfrak{A}}_1 = \mathfrak{A}_1$.

Let us assume now that the congruence $x^2 \equiv \mathbf{d} \pmod{4p}$ has no solutions. If $\mathfrak{A}$ is a complete module satisfying (5.35), then the least positive integer contained in $\mathfrak{A}$

is equal to p^δ with $\delta = 1$ or $\delta = 2$, and $\mathfrak{A} = \{p^\delta, p^\delta\gamma\}$. If $a\gamma^2 + b\gamma + c = 0$, where $a, b, c \in \mathbb{Z}$, $a > 0$, and $\gcd(a,b,c) = 1$, then by Lemma 5.19, we conclude that $p^2 = N(\mathfrak{A})N(p^\delta)N(\{1,\gamma\}) = p^{2\delta}/a$, $O(\mathfrak{A}) = \{1, a\gamma\}$, and $\delta(O(\mathfrak{A})) = b^2 - 4ac$. If $\delta = 2$, then $a = p^2$ and $\mathbf{d}l^2 = b^2 - 4ac \equiv b^2 \pmod{4p^2}$, which, since $p \nmid l$, contradicts the assumption. Thus $\delta = 1$, $a = 1$, $O(\mathfrak{A}) = O_l = \{1, \gamma\}$, and $\mathfrak{A} = p\{1,\gamma\} = pO_l$. On the other hand, the module pO_l clearly satisfies (5.35).

Every ideal $\mathfrak{A} \subset O_l$ with prime norm is clearly a prime ideal. Let $N(\mathfrak{A}) = p^2$. If $\alpha \in O(\mathfrak{A}) = O_l$ and $\alpha \notin \mathfrak{A}$, then the module $\mathfrak{A}' = \alpha O_l + \mathfrak{A}$ is complete and $N(\mathfrak{A}) = p$ or 1, since $\mathfrak{A}' \neq \mathfrak{A}$. By the above, the equality $N(\mathfrak{A}') = p$ in this case is impossible; hence $N(\mathfrak{A}') = 1$, $\mathfrak{A}' = O_l$, and α is invertible modulo $\mathfrak{A}$. Thus, the quotient $O_l/\mathfrak{A}$ is a field, and the ideal $\mathfrak{A}$ is prime. □

In view the above proposition it will be convenient to define the *sign* $\varepsilon_p(\delta)$ *of* δ *modulo a prime p* by

$$\varepsilon_p(\mathbf{d}) = \begin{cases} 1 & \text{if the congruence } x^2 \equiv \mathbf{d} \pmod{4p} \text{ is solvable and } p \nmid \mathbf{d}, \\ 0 & \text{if the congruence } x^2 \equiv \mathbf{d} \pmod{4p} \text{ is solvable and } p \mid \mathbf{d}, \\ -1 & \text{if the congruence } x^2 \equiv \mathbf{d} \pmod{4p} \text{ has no solutions.} \end{cases} \tag{5.36}$$

We conclude the survey part of this subsection with a description of the correspondence between modules in quadratic fields of negative discriminant and integral positive definite binary quadratic forms. For the quadratic forms, their matrices, and related notion, we shall use the notation (3.25)–(3.27) and the definitions of the beginning of this section. Let $K = \mathbb{Q}(\sqrt{\mathbf{d}})$ be an imaginary quadratic field with discriminant $\mathbf{d} < 0$, and let $\mathfrak{A}$ be a full module in K with the ring of multipliers of the form $O(\mathfrak{A}) = O_l$. With every $\mathbb{Z}$ basis α, β of $\mathfrak{A}$ ordered by the condition $\Im(\beta/\alpha) > 0$ we associate the binary quadratic form

$$\mathbf{a} = \mathbf{a}(\mathfrak{A}) = \frac{1}{N(\mathfrak{A})}(\alpha x + \beta y)(\overline{\alpha} x + \overline{\beta} y) = ax^2 + bxy + cy^2.$$

The form $\mathbf{a}$ is clearly positive definite. It follows easily from Lemma 5.1 that the form is integral and primitive with discriminant $\delta(\mathbf{a}) = b^2 - 4ac = \mathbf{d}l^2$ equal to the discriminant of the ring of multipliers $O(\mathfrak{A})$. Conversely, if $\mathbf{a}(X) = ax^2 + bxy + cy^2$ is a positive definite integral primitive quadratic form of discriminant $b^2 - 4ac = \mathbf{d}l^2$, then the module

$$\mathfrak{A} = \mathfrak{A}(\mathbf{a}) = \left\{ a, \frac{-b + l\sqrt{\mathbf{d}}}{2} \right\}$$

is a full module in K with ring of multipliers O_l contained in the ring. It is not difficult to check that the indicated correspondences define a bijection between the set of all classes of equivalent full modules in K with the ring of multipliers O_l and the set of all classes of properly equivalent positive definite integral primitive binary quadratic forms of discriminant $\mathbf{d}l^2$.

Let $\mathbf{a}$ and $\mathbf{a}'$ be two integral primitive positive definite binary quadratic forms of the same negative discriminant δ, and let A and A' be the matrices of the forms. Since the modules $\mathfrak{A} = \mathfrak{A}(\mathbf{a})$ and $\mathfrak{A}' = \mathfrak{A}(\mathbf{a}')$ are full modules in the field $K = \mathbb{Q}(\sqrt{\delta})$

with the same ring of multipliers of discriminant $\delta = \mathbf{d}l^2$, the same is true for their product $\mathfrak{A}\mathfrak{A}'$. We define the *Gaussian composition* $\mathbf{a} \times \mathbf{a}'$ *of the forms* $\mathbf{a}$ *and* $\mathbf{a}'$ as the form corresponding to the product $\mathfrak{A}\mathfrak{A}'$,

$$\mathbf{a} \times \mathbf{a}' = \mathbf{a}(\mathfrak{A}\mathfrak{A}'),$$

and denote by $A \times A'$ the matrix of $\mathbf{a} \times \mathbf{a}'$. For any full module $\mathfrak{A}$ in K with the same ring of multipliers, we denote by $A \times \mathfrak{A}$ the matrix of the quadratic form $\mathbf{a} \times \mathbf{a}(\mathfrak{A})$,

$$A \times \mathfrak{A} = A \times A(\mathfrak{A}), \tag{5.37}$$

where $A(\mathfrak{A})$ is the matrix of the form $\mathbf{a}(\mathfrak{A})$. The proper equivalence classes of the products depend only on proper classes of the quadratic forms or their matrices and the equivalence class of $\mathfrak{A}$. Thus, the above formulas define the products of corresponding classes.

Since the set of equivalence classes of full modules with the ring of multipliers of discriminant δ forms an abelian group under multiplication of the classes, the set of classes with respect to proper equivalence of integral primitive positive definite binary quadratic forms of discriminant δ and the set of proper classes of their matrices can also be endowed with a group structure. We shall denote by $H(\delta)$ all three of these naturally isomorphic class groups and by $h(\delta) = \#\{H(\delta)\}$ the *class number.*

We can now return to the action of Π-operators. First we shall examine the summation conditions in the formula (4.27) defining Π-operators. By Lemma 3.3, one can take

$$\Lambda_+\backslash\Lambda_+\mathrm{diag}(1,d)\Lambda_+ = \left\{\mathrm{diag}(1,d)U \,\middle|\, U \in \Lambda_d\backslash\Lambda_+\right\},$$

where

$$\Lambda_d = \Lambda_+ \bigcap \mathrm{diag}(1,d)^{-1}\Lambda_+\mathrm{diag}(1,d) = \left\{\begin{pmatrix}\alpha & \beta\\ \gamma & \delta\end{pmatrix} \in \Lambda_+ \,\middle|\, \beta \equiv 0 \pmod{d}\right\}.$$

It is easy to check that two matrices of Λ_+ with first rows (u_1, u_2) and (u_1', u_2') belong to the same left coset modulo Λ_d if and only if

$$(u_1, u_2) \cong (u_1', u_2') \pmod{d} \Longleftrightarrow u_1u_2' - u_2u_1' \equiv 0 \pmod{d}. \tag{5.38}$$

Thus, we have proved the following lemma.

Lemma 5.21. *One can take*

$$\Lambda_+\backslash\Lambda_+\mathrm{diag}(1,d)\Lambda_+$$
$$= \left\{\mathrm{diag}(1,d)U \,\middle|\, U = \begin{pmatrix}u_1 & u_2\\ v_1 & v_2\end{pmatrix}, (u_1, u_2) \in \mathbb{P}^1(\mathbb{Z}/d\mathbb{Z})\right\}, \tag{5.39}$$

where $\mathbb{P}^1(\mathbb{Z}/d\mathbb{Z})$ *is the "projective line modulo d", i.e., a set of representatives for classes of all coprime pairs of integers* (u_1, u_2) *modulo the equivalence* (5.38).

Exercise 5.22. Prove that the cardinal numbers $\kappa(d) = \#(\mathbb{P}^1(\mathbb{Z}/d\mathbb{Z}))$ satisfy the following rules: $\kappa(dd') = \kappa(d)\kappa(d')$ if $\gcd(d,d') = 1$, and $\kappa(p^\delta) = p^{\delta-1}(p+1)$ if p is a prime number.

Lemma 5.21 will allow us to rewrite formula (4.27) for the action of Hecke operators $|_{k,\chi}\Pi(d)$ on Fourier coefficients of functions of $\mathfrak{F}_\varepsilon$ in terms of points of the projective line modulo d. Let $\mathbf{a} = \mathbf{a}(\mathbf{u})$ with $\mathbf{u} = (u_1, u_2)$ be the quadratic form (5.25) with the matrix A, and $\mathbf{b}(\mathbf{u},\mathbf{v}) = \mathbf{u}A\,{}^t\mathbf{v} = \mathbf{b}(\mathbf{v},\mathbf{u})$ the corresponding bilinear form. Then for D of the form

$$D = \begin{pmatrix} 1 & 0 \\ 0 & d \end{pmatrix}\begin{pmatrix} u_1 & u_2 \\ v_1 & v_2 \end{pmatrix} = \begin{pmatrix} \mathbf{u} \\ d\mathbf{v} \end{pmatrix},$$

we have

$$A[{}^tD] = DA\,{}^tD = \begin{pmatrix} \mathbf{u}A\,{}^t\mathbf{u} & d\mathbf{u}A\,{}^t\mathbf{v} \\ d\mathbf{v}A\,{}^t\mathbf{u} & d^2\mathbf{v}A\,{}^t\mathbf{v} \end{pmatrix} = \begin{pmatrix} 2\mathbf{a}(\mathbf{u}) & d\mathbf{b}(\mathbf{u},\mathbf{v}) \\ d\mathbf{b}(\mathbf{v},\mathbf{u}) & 2d^2\mathbf{a}(\mathbf{v}) \end{pmatrix}.$$

By Lemma 5.21, in this notation we can rewrite formula (4.27) in the form

$$(f|_{k,\chi}\Pi(d))(A) = \chi(d)d^{k-2}(f|P(d))(A), \tag{5.40}$$

where

$$(f|P(d))(A) = \sum_{\substack{\mathbf{u}\in\mathbb{P}^1(\mathbb{Z}/d\mathbb{Z}),\\ \mathbf{a}(\mathbf{u})\equiv 0 \pmod{d}}} f\left(\begin{pmatrix} 2\mathbf{a}(\mathbf{u})/d & \mathbf{b}(\mathbf{u},\mathbf{v}) \\ \mathbf{b}(\mathbf{v},\mathbf{u}) & 2d\mathbf{a}(\mathbf{v}) \end{pmatrix}\right), \tag{5.41}$$

and where for each $\mathbf{u} \in \mathbb{P}^1(\mathbb{Z}/d\mathbb{Z})$, by $\mathbf{v}$ is denoted a 2-row such that $\binom{\mathbf{u}}{\mathbf{v}} \in \Lambda_+$. Since the definition of operators $|P(d)$ is independent of the choice of k and χ, according to (5.40) we can express the operators $|P(d)$ on Fourier coefficients of functions of $\mathfrak{F}_\varepsilon$ through the operators $|_{k,\chi}\Pi(d)$, for arbitrary k and χ satisfying $(-1)^k\chi(-1) = \varepsilon(-1)$ and $\chi(d) \neq 0$, by the formula

$$|P(d) = \overline{\chi}(d)d^{2-k}|_{k,\chi}\Pi(d). \tag{5.42}$$

The next key lemma interprets values of the images $(f|P(p))(A)$ for prime p and primitive A in terms of multiplication of quadratic modules.

Lemma 5.23. *Let $A = \begin{pmatrix} 2a & b \\ b & 2c \end{pmatrix}$ be the matrix of an integral primitive quadratic form $\mathbf{a}$ of negative discriminant $\delta = b^2 - 4ac = \mathbf{d}l^2$, where $\mathbf{d}$ is the discriminant of the imaginary quadratic field $\mathrm{K} = \mathbb{Q}(\sqrt{\delta})$. Then for each function f on $\mathbb{E}^2$ depending only on classes of even matrices relative to proper equivalence and each prime number p not dividing the number l, we have*

$$(f|P(p))(A) = \begin{cases} f(A \times \mathfrak{P}) + f(A \times \overline{\mathfrak{P}}) & \text{if } \varepsilon_p(\mathbf{d}) = 1, \\ f(A \times \mathfrak{P}) & \text{if } \varepsilon_p(\mathbf{d}) = 0, \\ 0 & \text{if } \varepsilon_p(\mathbf{d}) = -1, \end{cases}$$

where $\varepsilon_p(\mathbf{d})$ is the sign (5.36), $\mathfrak{P}$ a regular prime ideal of norm p of the order $O_l \subset K$ with discriminant $\mathbf{d}l^2$, $\overline{\mathfrak{P}}$ the conjugate ideal, and where $\times$ is the multiplication (5.37).

Proof. Let us consider the congruence

$$\mathbf{a}(x_1,x_2) = ax_1^2 + bx_1x_2 + cx_2^2 \equiv 0 \pmod{p}. \tag{5.43}$$

It is easy to check in the framework of the elementary theory of quadratic congruences that for each prime p not dividing l, the number of different points of the projective line $\mathbb{P}^1(\mathbb{Z}/p\mathbb{Z})$ modulo p, i.e., the number of inequivalent coprime pairs of integers with respect to the equivalence (5.38) modulo p, satisfying this congruence is equal to $1+\varepsilon_p(\mathbf{d})$.

Let $\varepsilon_p(\mathbf{d}) = 1$, and let $\mathbf{u} = (u_1,u_2)$ and $\mathbf{u}' = (u_1',u_2')$ be two inequivalent modulo p primitive solutions of (5.43). Since the discriminant $\delta = b^2 - 4ac$ is not divisible by p, we may assume that $\mathbf{a}(\mathbf{u}) = pa_1$, $\mathbf{a}(\mathbf{u}') = pa_2$, where a_1 and a_2 are not divisible by p. We can find integers v_1, v_2, v_1', v_2' such that

$$U_1 = \begin{pmatrix} u_1 & u_2 \\ v_1 & v_2 \end{pmatrix}, \quad U_2 = \begin{pmatrix} u_1' & u_2' \\ v_1' & v_2' \end{pmatrix} \in \Lambda_+.$$

We set

$$A_1 = U_1 A\, {}^tU_1 = \begin{pmatrix} 2pa_1 & b_1 \\ b_1 & c_1 \end{pmatrix}, \quad A_1' = \begin{pmatrix} 2a_1 & b_1 \\ b_1 & 2pc_1 \end{pmatrix},$$

$$A_2 = U_2 A\, {}^tU_2 = \begin{pmatrix} 2pa_2 & b_2 \\ b_2 & c_2 \end{pmatrix}, \quad A_2' = \begin{pmatrix} 2a_2 & b_2 \\ b_2 & 2pc_2 \end{pmatrix}.$$

Then by (5.41), we get $(P(p)f)(A) = f(A_1') + f(A_2')$, and it is sufficient to prove that with the appropriate choice of regular prime ideal $\mathfrak{P}$ of norm p in O_l, one has

$$A_1' \sim A \times \mathfrak{P} \quad \text{and} \quad A_2' \sim A \times \overline{\mathfrak{P}}, \tag{5.44}$$

where the symbol $\sim$ stands for proper equivalence. Let us consider the modules

$$\mathfrak{A} = \left\{ p, \frac{-b_1 + \sqrt{\delta}}{2} \right\}, \qquad \mathfrak{A}' = \left\{ p, \frac{-b_2 + \sqrt{\delta}}{2} \right\}$$

of the field K. It follows from Lemma 5.19 that the ring of multipliers of each of these modules is O_l, that they are both contained in O_l, and that they have the norm $N(\mathfrak{A}) = N(\mathfrak{A}') = p$. Let us show that $\mathfrak{A} + \mathfrak{A}' = O_l$. To prove this it is obviously sufficient to check that the number $(b_2 - b_1)/2$ is not divisible by p. We set

$$U_2 U_1^{-1} = T = \begin{pmatrix} t_1 & t_2 \\ t_3 & t_4 \end{pmatrix} \in \Lambda_+.$$

Since the solutions (u_1,u_2) and (u'_1,u'_2) are not equivalent modulo p, we have $t_2 = u_1u'_2 - u_2u'_1 \not\equiv 0 \pmod p$. It follows from the relation $TA_1{}^tT = A_2$ that $pa_1t_1^2 + b_1t_1t_2 + c_1t_2^2 = pa_2$ and $2pa_1t_1t_3 + b_1(t_1t_4 + t_2t_3) + 2c_1t_2t_4 = b_2$. From the first equality we obtain the congruence $b_1t_1 + c_1t_2 \equiv 0 \pmod p$, and it follows from the second that

$$\frac{b_2 - b_1}{2} \equiv \frac{b_1(t_1t_4 + t_2t_3) + 2c_1t_2t_4 - b_1}{2} = t_2(b_1t_3 + c_1t_4) \pmod p.$$

If $(b_2 - b_1)/2$ were divisible by p, we would get the system of congruences $b_1t_1 + c_1t_2 \equiv b_1t_3 + c_1t_4 \equiv 0 \pmod p$, which would imply that $b_1 \equiv c_1 \equiv 0 \pmod p$, a contradiction, because $\delta = b_1^2 - 4pa_1c_1$ is not divisible by p. It follows from what we have proved and Proposition 5.20 that one can set $\mathfrak{A}' = \mathfrak{P}$ and $\mathfrak{A} = \overline{\mathfrak{P}}$, where $\mathfrak{P}, \overline{\mathfrak{P}}$ are the unique pair of regular conjugate distinct prime ideals in O_l with norm p. In order to prove (5.43), let us denote by $\mathbf{a}_1, \mathbf{a}_2, \mathbf{a}'_1$, and $\mathbf{a}'_2$ the quadratic forms with matrices A_1, A_2, A'_1, and A'_2, respectively, and by $\mathfrak{A}_1, \mathfrak{A}_2, \mathfrak{A}'_1$, and $\mathfrak{A}'_2$ the corresponding modules in K. If we set $\gamma_1 = (-b_1 + \sqrt{\delta})/2$, then clearly $\gamma_1^2 = -pa_1c_1 - b_1\gamma$, and since a_1 and p are coprime, we obtain

$$\mathfrak{A}'_1\overline{\mathfrak{P}} = \{a_1, \gamma_1\}\{p, \gamma_1\} = \{pa_1, p\gamma_1, a_1\gamma_1, -pa_1c_1 - b_1\gamma_1\} = \{pa_1, \gamma_1\} = \mathfrak{A}_1.$$

Hence $\mathfrak{A}_1\mathfrak{P} = \mathfrak{A}'_1\mathfrak{P}\overline{\mathfrak{P}} = \mathfrak{A}'_1N(\mathfrak{P}) \sim \mathfrak{A}'_1$, and $A'_1 \sim A_1 \times \mathfrak{P} \sim A \times \mathfrak{P}$. Similarly, $A'_2 \sim A \times \overline{\mathfrak{P}}$. This proves (5.44) and the first formula of the lemma.

Now let $\varepsilon_p(d) = 0$ and let $\mathbf{u} = (u_1, u_2)$ be the unique solution of (5.43) modulo equivalence (5.38). We can find integers v_1, v_2 such that $U_1 = \left(\begin{smallmatrix} u_1 & u_2 \\ v_1 & v_2 \end{smallmatrix}\right) \in \Lambda_+$ and set

$$A_1 = U_1A{}^tU_1 = \begin{pmatrix} 2pa_1 & b_1 \\ b_1 & c_1 \end{pmatrix}, \quad A'_1 = \begin{pmatrix} 2a_1 & b_1 \\ b_1 & 2pc_1 \end{pmatrix}.$$

Then the right-hand side of (5.41) is equal to $f(A'_1)$, and it is sufficient to prove that

$$A'_1 \sim A \times \mathfrak{P}, \tag{5.45}$$

where $\mathfrak{P}$ is the unique regular prime ideal of norm p in O_l. By our assumption, the discriminant $\delta = \mathbf{d}l^2 = b^2 - 4ac = b_1^2 - 4pa_1c_1$ is divisible by p and $p \nmid l$. It follows that

$$b_1 \equiv 0 \pmod p \quad \text{and} \quad \mathbf{d}l^2 \equiv -4pa_1c_1 \pmod{p^2}.$$

If $p \neq 2$, since $\mathbf{d}$ is not divisible by p^2, the last congruence implies that a_1c_1 is not divisible by p. If $p = 2$, then $\mathbf{d} = 4d_0$ with $d_0 \equiv 2, 3 \pmod 4$. Let us set $b_1 = 2b_0$. Then $\delta = 4d_0l^2 = 4(b_0^2 - 2a_1c_1)$, whence

$$2a_1c_1 = b_0^2 - d_0l^2 \equiv b_0^2 - d_0 \pmod 4.$$

The number $b_0^2 - d_0$ is not divisible by 4, since $b_0^2 \equiv 0, 1 \pmod 4$ and $d_0 \equiv 3, 3 \pmod 4$. Hence, a_1c_1 is not divisible by p also for $p = 2$. Further, since the quadratic form $\mathbf{a}_1$ with matrix A_1 together with $\mathbf{a}$ is primitive, and $p \nmid a_1$, it follows that the form $\mathbf{a}'_1$ with matrix A'_1 is also primitive. Let us consider the module $\mathfrak{A} = \{p, \gamma\}$, where $\gamma = (-b_1 + \sqrt{\delta})/2$ satisfies $\gamma^2 + b_1\gamma + pa_1c_1 = 0$. By Lemma 5.19 and Proposition 5.20, the module $\mathfrak{A} = \mathfrak{P}$ is the unique regular prime ideal of O_l of norm p. Let us denote by $\mathbf{a}_1$ and $\mathbf{a}_1'$ the quadratic forms with matrices A_1 and A_1', respectively, and by $\mathfrak{A}_1$ and $\mathfrak{A}_1'$ the corresponding modules in K. Since $\gamma^2 = -b_1\gamma - pa_1c_1$ and $p \nmid a_1$, we have

$$\mathfrak{A}_1'\mathfrak{P} = \{a_1, \gamma\}\{p, \gamma\} = \{pa_1, p\gamma, a_1\gamma, -pa_1c_1 - b_1\gamma_1\} = \{pa_1, \gamma\} = \mathfrak{A}_1.$$

Hence, since $\mathfrak{P} = \overline{\mathfrak{P}}$, we have $\mathfrak{A}_1\mathfrak{P} = \mathfrak{A}_1'\mathfrak{P}\overline{\mathfrak{P}} \sim \mathfrak{A}_1'$, and the corresponding matrices satisfy (5.45).

Finally, if $\varepsilon_p(d) = -1$, then, as was noted at the beginning of the proof, the congruence (5.43) has no solutions on the projective line modulo p, which proves the last formula of the lemma. □

Exercise 5.24. In the notation of Lemma 5.23, assuming that $\varepsilon_p(\mathbf{d}) = 0$ and $p|l$, prove the formula $(f|P(p))(A) = f(A \times O_{l/p})$.

Along with operators corresponding to elements $\Pi(p)$ we shall also need operators associated with elements $\Psi(p) \in \mathcal{L}_p^2$ defined by formula (3.145). The action of these operators can be naturally reduced to computation of simple trigonometric sums, which will be done in the next lemma.

Lemma 5.25. *In the notation and under the assumptions of the previous lemma and Lemma 3.68, the following formulas hold*

$$S_p(A) = \sum_{\substack{B = {}^tB \in \mathbb{Z}_2^2/p\mathbb{Z}_2^2, \\ r_p(B)=1}} e^{\pi i \sigma(AB)/p} = \begin{cases} p-1 & \text{if } \varepsilon_p(\mathbf{d}) = 1, \\ -1 & \text{if } \varepsilon_p(\mathbf{d}) = 0, \\ -(p+1) & \text{if } \varepsilon_p(\mathbf{d}) = -1, \end{cases} \tag{5.46}$$

Proof. The set of matrices of the form

$$\left\{ \begin{pmatrix} \alpha & 0 \\ 0 & 0 \end{pmatrix}, \begin{pmatrix} \gamma\beta^2 & \gamma\beta \\ \gamma\beta & \gamma \end{pmatrix} \;\middle|\; \alpha, \gamma = 1, \dots, p-1, \beta = 0, 1, \dots, p-1 \right\}$$

can clearly be taken as a summation set in the sum $S_p(A)$. Thus, since $A = \left(\begin{smallmatrix} 2a & b \\ b & 2c \end{smallmatrix}\right)$, we obtain

$$S_p(A) = \sum_{\alpha} e^{2\pi i a\alpha/p} + \sum_{\gamma,\beta} e^{2\pi i \gamma(a\beta^2 + b\beta + c)/p}.$$

The sum on α is obviously equal to $p-1$ if $p|a$ and -1 if $p \nmid a$; similarly, the sum on γ is $p-1$ if $p|a\beta^2 + b\beta + c$ and -1 if $p \nmid a\beta^2 + b\beta + c$. Hence if, for example, $\varepsilon_p(\mathbf{d}) = -1$, then the second of the alternatives is realized in all cases, and we have $S_p(A) = -1 + p(-1) = -(p+1)$. The other cases are similar, and we leave their consideration to the reader as a useful exercise on quadratic congruences. □

The above lemmas together with Lemma 3.71 will allow us to discover the action of the operators $|P(d)$ with composite d on Fourier coefficients of functions of $\mathfrak{F}_\varepsilon$ in terms of characters of class groups.

Theorem 5.26. *Let δ be a negative integer written in the form $\delta = \mathbf{d}l^2$, where $\mathbf{d}$ is the discriminant of the imaginary quadratic field $K = \mathbb{Q}(\sqrt{\delta})$ and l is a positive integer. Let O_l be the order in K of the discriminant δ, $A_1, \dots, A_h$ with $h = h(\delta)$ a full system of representatives of classes of properly equivalent even positive definite primitive matrices of order two and the discriminant δ, and let ψ be a character of the class group $H(\delta)$. Then for every function f on $\mathbb{E}^2$ whose values are Fourier coefficients of a function of $\mathfrak{F}_\varepsilon$ and positive integer d prime to l, the following formulas hold for the action of operators* (5.41) *on averaging*

$$\psi(A_1)f(A_1) + \cdots + \psi(A_h)f(A_h)$$

of f over the group $H(\delta)$ with character ψ:

$$\left(\sum_{i=1}^{h} \psi(A_i)f(A_i)\right)|P(d) \quad \left(= \sum_{i=1}^{h} \psi(A_i)(f|P(d))(A_i)\right)$$
$$= \lambda(P(d);\, \psi, \delta)\left(\sum_{i=1}^{h} \psi(A_i)f(A_i)\right), \tag{5.47}$$

where the numbers $\lambda(P(d);\, \psi, \delta)$ satisfy the formal identity

$$\sum_{d\in\mathbb{N},\, \gcd(d,l)=1} \frac{\lambda(P(d);\, \psi, \delta)}{d^s} = \prod_{p\in\mathbb{P},\, p\nmid l} \left(1 - \frac{1}{p^{2s}}\right) \prod_{\mathfrak{P}\,|\,p} \left(1 - \frac{\psi(\mathfrak{P})}{N(\mathfrak{P})^s}\right)^{-1}, \tag{5.48}$$

where $\mathfrak{P}$ in the inner product on the right runs through all regular prime ideals of the ring O_l dividing the principal ideal pO_l.

Proof. Since the function f depends only on proper classes of even matrices, by Lemma 5.23 in the notation of the lemma, for a prime p not dividing l, we obtain

$$\left(\sum_{i=1}^{h} \psi(A_i)f(A_i)\right)|P(p) = \sum_{i=1}^{h} \psi(A_i)(f|P(p))(A_i)$$
$$= \begin{cases} \sum_{i=1}^{h} \psi(A_i)\big(f(A_i \times \mathfrak{P}) + f(A_i \times \overline{\mathfrak{P}})\big) & \text{if } \varepsilon_p(\mathbf{d}) = 1, \\ \sum_{i=1}^{h} \psi(A_i)f(A_i \times \mathfrak{P}) & \text{if } \varepsilon_p(\mathbf{d}) = 0, \\ 0 & \text{if } \varepsilon_p(\mathbf{d}) = -1, \end{cases}$$

which clearly implies the formula (5.47) with $d = p$ and

$$\lambda(P(p);\, \psi, \delta) = \begin{cases} \psi(\mathfrak{P}) + \psi(\overline{\mathfrak{P}}) & \text{if } \varepsilon_p(\mathbf{d}) = 1, \\ \psi(\overline{\mathfrak{P}}) = \psi(\mathfrak{P}) & \text{if } \varepsilon_p(\mathbf{d}) = 0, \\ 0 & \text{if } \varepsilon_p(\mathbf{d}) = -1. \end{cases} \tag{5.49}$$

In order to reduce the action of operators $|P(d)$ for composite d to the cases of prime divisors of d, we shall use the relation (5.42) between operators $|P(d)$ and

$|_k\Pi = |_{k,1}\Pi(d)$ with $\chi = 1$ (the unit character modulo 1) and relations for elements $\Pi(d)$ obtained in Section 3.5. First of all, by Lemma 3.62, we have

$$\begin{aligned}|P(d)|P(d_1) &= d^{k-2}d_1^{k-2}|\Pi(d)|\Pi(d_1)\\ &= (dd_1)^{k-2}|\Pi(dd_1) = |P(dd_1) \text{ if } \gcd(d,d_1)=1. \end{aligned} \tag{5.50}$$

Then, by Lemma 3.71, for each prime p we get the following summation formula for the formal power series with operators $|P(p^\nu)$ as coefficients:

$$\begin{aligned}\sum_{\nu\ge 0}|P(p^\nu)v^\nu &= \sum_{\nu\ge 0}|_k\Pi(p^\nu)(p^{2-k}v)^\nu\\ &= (1-|_kp^2\langle\mathbf{p}\rangle_2(p^{2-k}v)^2)\\ &\quad\times\left(1-|_k\Pi(p)p^{2-k}v+(|_kp\langle\mathbf{p}\rangle_2+|_kp\Psi(p))(p^{2-k}v)^2\right)^{-1}\\ &= (1-|_k\langle\mathbf{p}\rangle_2p^{6-2k}v^2)\left(1-|P(p)v+(|_k\langle\mathbf{p}\rangle_2+|_k\Psi(p))p^{5-2k}v^2\right)^{-1}. \end{aligned} \tag{5.51}$$

Let us consider now the action of operators $|_k\langle\mathbf{p}\rangle_2$ and $|_k\Psi(p_i)$ on the average. By formula (4.26) for $n=2$ and $\chi=1$, we have

$$\left(\sum_{i=1}^h\psi(A_i)f(A_i)\right)|_k\langle\mathbf{p}\rangle_2 = p^{2k-6}\left(\sum_{i=1}^h\psi(A_i)f(A_i)\right). \tag{5.52}$$

By formula (4.22) of Lemma 4.13 for $\chi=1$ applied to the double cosets entering in the decomposition (3.45) of the element $\Psi(p)$, for each $A\in\mathbb{E}^2$, one can write $(f|_k\Psi(p))(A) = p^{2k-6}S_p(A)f(A)$, where $S_p(A)$ is the trigonometric sum (5.46). Hence, by Lemma 5.25, we derive the formula

$$\left(\sum_{i=1}^h\psi(A_i)f(A_i)\right)|_k\Psi(p) = p^{2k-6}\lambda(\Psi(p);\psi,\delta)\left(\sum_{i=1}^h\psi(A_i)f(A_i)\right), \tag{5.53}$$

where

$$\lambda(\Psi(p);\psi,\delta) = \begin{cases}(p-1) & \text{if } \varepsilon_p(\mathbf{d})=1,\\ -1 & \text{if } \varepsilon_p(\mathbf{d})=0,\\ -(p+1) & \text{if } \varepsilon_p(\mathbf{d})=-1.\end{cases}$$

It follows from the relations (5.50) and (5.51) that each of the operators $|P(d)$ with $d=\prod_i p_i^{\nu_i}$, where p_i are different prime divisors of d, is a polynomial in operators $|P(p_i)$, $|_k\langle\mathbf{p}_i\rangle_2$, and $|_k\Psi(p_i)$ that commute with each other. Then the relations (5.47) with some coefficients $\lambda(P(d);\psi,\delta)$ follow for every d prime to l from the formulas for primes proved above. In order to compute the coefficients, we note first that by (5.50), we have

$$\lambda(P(dd_1);\psi,\delta) = \lambda(P(d);\psi,\delta)\lambda(P(d_1);\psi,\delta) \quad \text{if } \gcd(d,d_1)=1, \tag{5.54}$$

which implies the formal Euler product factorization

$$\sum_{d\in\mathbb{N},\,\gcd(d,l)=1} \frac{\lambda(P(d);\psi,\delta)}{d^s} = \prod_{p\in\mathbb{P},\,p\nmid l} \left(\sum_{\nu\geq 0} \lambda(P(p^\nu);\psi,\delta)\frac{1}{p^{\nu s}} \right). \tag{5.55}$$

It follows from the summation formula (5.51) and formulas (5.47) with $d = p$, (5.52), and (5.53) that each of the p-factors of the last Euler product can be summed in the form

$$\sum_{\nu\geq 0} \lambda(P(p^\nu);\psi,\delta)\frac{1}{p^{\nu s}} = \left(1-\frac{1}{p^{2s}}\right) \times \left(1-\frac{\lambda(P(p);\psi,\delta)}{p^s}+\frac{(1+\lambda(\Psi(p);\psi,\delta))p^{-1}}{p^{2s}}\right)^{-1}.$$

By substituting here the values for $\lambda(P(p);\psi,\delta)$ and $\lambda(|_k\Psi(p);\psi,\delta)$ computed above, we obtain that the fraction on the right is equal to

$$\left(1-\frac{1}{p^{2s}}\right)\left(1-\frac{(\psi(\mathfrak{P})+\psi(\overline{\mathfrak{P}}))}{p^s}+\frac{(1+p-1)p^{-1}}{p^{2s}}\right)^{-1} = \left(1-\frac{1}{p^{2s}}\right)\left(\left(1-\frac{\psi(\mathfrak{P})}{N(\mathfrak{P})^s}\right)\left(1-\frac{\psi(\overline{\mathfrak{P}})}{N(\overline{\mathfrak{P}})^s}\right)\right)^{-1}$$

if $\varepsilon_p(\mathbf{d}) = 1$, where $\mathfrak{P}$ and $\overline{\mathfrak{P}}$ is a pair of conjugate regular prime ideals of O_l of norm p,

$$\left(1-\frac{1}{p^{2s}}\right)\left(1-\frac{\psi(\mathfrak{P})}{p^s}+\frac{(1-1)p^{-1}}{p^{2s}}\right)^{-1} = \left(1-\frac{1}{p^{2s}}\right)\left(1-\frac{\psi(\mathfrak{P})}{N(\mathfrak{P})^s}\right)^{-1}$$

if $\varepsilon_p(\mathbf{d}) = 0$, where $\mathfrak{P}$ is a regular prime ideal of O_l of norm p, and

$$\left(1-\frac{1}{p^{2s}}\right)\left(1+\frac{(1-(p+1))p^{-1}}{p^{2s}}\right)^{-1} = \left(1-\frac{1}{p^{2s}}\right)\left(1-\frac{\psi(\mathfrak{P})}{N(\mathfrak{P})^s}\right)^{-1}$$

if $\varepsilon_p(\mathbf{d}) = -1$, where $\mathfrak{P} = pO_l$ is a regular prime ideal of O_l of norm p^2 (with $\psi(\mathfrak{P}) = 1$). Hence, according to Proposition 5.20, we conclude that in all three cases, the p-factor of the factorization (5.55) can be written in the form

$$\sum_{\nu\geq 0} \lambda(P(p^\nu);\psi,\delta)\frac{1}{p^{\nu s}} = \left(1-\frac{1}{p^{2s}}\right)\prod_{\mathfrak{P}|p}\left(1-\frac{\psi(\mathfrak{P})}{N(\mathfrak{P})^s}\right)^{-1}, \tag{5.56}$$

where $\mathfrak{P}$ ranges over all regular prime ideals of the ring O_l dividing the principal ideal pO_l. The formal Euler factorization (5.48) follows from (5.55) and (5.56). □

Exercise 5.27. Let $\rho(d,A)$ be the number (5.31) of solutions of the congruence $\mathbf{a}(\mathbf{u}) \equiv 0 \pmod{d}$, where $\mathbf{a}$ is the quadratic form with matrix A, on the projective line modulo d. In the notation and under the assumptions of Theorem 5.26, prove the formal identity

$$\sum_{d\in\mathbb{N},\,\gcd(d,l)=1} \frac{\rho(d,A)}{d^s} = \prod_{p\in\mathbb{P},\,p\nmid l}\left(1-\frac{1}{p^{2s}}\right)\prod_{\mathfrak{P}\,|\,p}\left(1-\frac{1}{N(\mathfrak{P})^s}\right)^{-1}.$$

Multiplicative Properties of Fourier Coefficients. We return to a nonzero modular form F of weight $k>0$ and character χ modulo q for the group $\Gamma_0^2(q)$ with Fourier expansion (5.24) and assume that F is an eigenfunction for Hecke operators $|T(m)=|_{k,\chi}T(m)$ corresponding to all elements $T(m)\in\mathcal{L}_0(q)=\mathcal{L}_0^2(q)$ of the form (3.82), where $K=\Gamma_0^2(q)$, with eigenvalues $\lambda(m)=\lambda_F(m)$:

$$F|T(m)=\lambda(m)F \quad \text{for all } m \text{ prime to } q. \tag{5.57}$$

Then by Corollary 5.17, for every primitive matrix $A\in\mathbb{E}^2$ and positive integer ν coprime to q, the Fourier coefficients of F and the eigenvalues are linked by the relation (5.30), which we shall rewrite in the form

$$\lambda(m)f(\nu A) = \sum_{d,d_1\in\mathbb{N},\,dd_1=m} \chi(d_1)d_1^{k-2}((f|\Pi_+(\nu d))|P(d_1))(A), \tag{5.58}$$

where $|\Pi_+=|_{k,\chi}\Pi_+$ are the operators (4.25) and the operators P are defined by (5.41). Let, as in Theorem 5.26, δ be a negative integer written in the form $\delta=\mathbf{d}l^2$, where $\mathbf{d}$ is the discriminant of the imaginary quadratic field $K=\mathbb{Q}(\sqrt{\delta})$ and l is a positive integer. Let O_l be the order in K of discriminant δ, $A_1,\dots,A_h$ with $h=h(\delta)$ a full system of representatives of classes of properly equivalent even positive definite primitive matrices of order two with discriminant δ, and let ψ be a character of the class group $H(\delta)$. Let us multiply both sides of the relation (5.58) with $A=A_i$ by $\psi(A_i)$ and sum these relations over $i=1,\dots,h$. We get the relation

$$\lambda(m)\sum_{i=1}^{h}\psi(A_i)f(\nu A_i) = \sum_{\substack{d,\ d_1\in\mathbb{N},\\ dd_1=m}} \chi(d_1)d_1^{k-2}\sum_{i=1}^{h}\psi(A_i)((f|\Pi_+(d\nu))|P(d_1))(A_i).$$

If m is also coprime to l, then, since values of the functions $f|\Pi_+(\nu d)$ together with values of f can be considered as Fourier coefficients of functions of $\mathfrak{F}_\varepsilon$, we can apply relations (5.47) of Theorem 5.26 to the inner sums on the right of the last formula, which leads us the relations

$$\begin{aligned}
&\lambda(m)\sum_{i=1}^{h}\psi(A_i)f(\nu A_i)\\
&\quad= \sum_{d,\ d_1\in\mathbb{N},\,dd_1=m} \chi(d_1)d_1^{k-2}\lambda(P(d_1);\,\psi,\delta)\sum_{i=1}^{h}\psi(A_i)(f|\Pi_+(d\nu)))(A_i)\\
&\quad= \sum_{d,\ d_1\in\mathbb{N},\,dd_1=m} \chi(d_1)d_1^{k-2}\lambda(P(d_1);\,\psi,\delta)\sum_{i=1}^{h}\psi(A_i)f(d\nu A_i).
\end{aligned}$$

Hence, dividing both parts by $m^s = d^s d_1^s$ and summing over all $m \in \mathbb{N}$ with $\gcd(m, lq) = 1$, we come to the following generalization of the relations (5.32), obtained under the assumption that $h(\delta) = 1$, to the case of arbitrary class number:

$$\sum_{i=1}^{h} \psi(A_i) f(\nu A_i) \sum_{m \in \mathbb{N}, \gcd(m, lq)=1} \frac{\lambda(m)}{m^s}$$
$$= \sum_{d_1 \in \mathbb{N}, \gcd(d_1, lq)=1} \frac{\chi(d_1) d_1^{k-2} \lambda(P(d_1); \psi, \delta)}{d_1^s} \sum_{i=1}^{h} \psi(A_i) \sum_{d \in \mathbb{N}, \gcd(d, lq)=1} \frac{f(d \nu A_i)}{d^s}, \tag{5.59}$$

where ν is an arbitrary positive integer prime to all integers $m \in \mathbb{N}$ satisfying $\gcd(m, lq) = 1$, in other words, ν satisfies $\nu | (lq)^\infty$, i.e., divides a power of lq. Ultimately, we come to the following theorem on multiplicative properties of Fourier coefficients of eigenfunctions.

Theorem 5.28. *Let F be a nonzero modular form of weight $k \in \mathbb{N}$ and character χ modulo q for the group $\Gamma_0^2(q)$. Suppose that F is an eigenfunction satisfying (5.57). Let δ be a negative integer, $\delta = \mathbf{d} l^2$, where $\mathbf{d}$ is the discriminant of the field $\mathrm{K} = \mathbb{Q}(\sqrt{\delta})$ and $l \in \mathbb{N}$, $A_1, \ldots, A_h$ with $h = h(\delta)$ a full system of representatives of classes of properly equivalent even positive definite primitive matrices of order two with discriminant δ, and let ψ be a character of the group $H(\delta)$ of the classes of matrices with respect to the Gaussian composition (5.37). Then for each positive integer ν dividing a power of lq, the following formal identity holds:*

$$\sum_{i=1}^{h} \psi(A_i) f(\nu A_i) \prod_{p \in \mathbb{P}, p \nmid lq} Q_{p,F}^{-1}(p^{-s})$$
$$= \prod_{\mathfrak{P} \nmid lq} \left(1 - \frac{\psi(\mathfrak{P}) \chi(N(\mathfrak{P}))}{N(\mathfrak{P})^{s-k+2}}\right)^{-1} \sum_{i=1}^{h} \psi(A_i) \sum_{d \in \mathbb{N}, \gcd(d, lq)=1} \frac{f(d \nu A_i)}{d^s}, \tag{5.60}$$

where

$$Q_{p,F}(v) = 1 - \lambda(p) v + (\lambda(p)^2 - \lambda(p^2) - \chi(p^2) p^{2k-4}) v^2$$
$$- \chi(p^2) p^{2k-3} \lambda(p) v^3 + \chi(p^4) p^{4k-6} v^4, \tag{5.61}$$

and $\mathfrak{P}$ on the right ranges over all regular prime ideals of the ring O_l not dividing the principal ideal lqO_l.

Proof. By Theorem 3.46, we have

$$\lambda(mm') = \lambda(m)\lambda(m') \quad \text{if} \quad \gcd(m, m') = \gcd(mm', q) = 1.$$

By Proposition 3.53, for each prime number p not dividing q, we get the formal identity

$$\sum_{\beta=0}^{\infty} \lambda(p^\beta)v^\beta = \sum_{\beta=0}^{\infty} \lambda(T(p^\beta))v^\beta$$
$$= (1-p^2\lambda(\langle p\rangle)v^2)(1-\lambda(q_1(p))v+\lambda(q_2(p))v^2-\lambda(q_3(p))v^3+\lambda(q_4(p))v^4)^{-1},$$

where $\lambda(q_1(p)) = \lambda(T(p)) = \lambda(p)$,

$$\lambda(q_2(p)) = \lambda(T(p)^2) - \lambda(T(p^2)) - p^2\lambda(\langle p\rangle) = \lambda(p)^2 - \lambda(p^2) - p^2\lambda(\langle p\rangle),$$

$\lambda(q_3(p)) = p^3\lambda(\langle p\rangle)\lambda(p)$, and $\lambda(q_4(p)) = p^6\lambda(\langle p\rangle)^2$. Since by (4.26), we can write $\lambda(\langle p\rangle) = \lambda(\langle \mathbf{p}\rangle) = \chi(p^2)p^{2k-6}$, it follows that

$$\sum_{\beta=0}^{\infty} \lambda(p^\beta)v^\beta = (1-\chi(p^2)p^{2k-4}v^2)Q_{p,F}^{-1}(v),$$

where $Q_{p,F}(v)$ is the polynomial (5.61). Thus, we obtain the formal Euler product factorization

$$\sum_{m\in\mathbb{N},\,\gcd(m,lq)=1} \frac{\lambda(m)}{m^s} = \prod_{p\in\mathbb{P},\,p\nmid lq} \sum_{\beta=0}^{\infty} \frac{\lambda(p^\beta)}{p^{\beta s}}$$
$$= \prod_{p\in\mathbb{P},\,p\nmid lq} (1-\chi(p^2)p^{-2(s-k+2)})Q_{p,F}^{-1}(p^{-s}). \qquad (5.62)$$

On the other hand, by (5.48) we can write

$$\sum_{d\in\mathbb{N},\,\gcd(d,l)=1} \frac{\chi(d)d^{k-2}\lambda(P(d);\psi,\delta)}{d^s} = \sum_{d\in\mathbb{N},\,\gcd(d,l)=1} \lambda(P(d);\psi,\delta)\frac{\chi(d)}{d^{s-k+2}}$$
$$= \prod_{p\in\mathbb{P},\,p\nmid l} \left(1-\frac{\chi(p^2)}{p^{2(s-k+2)}}\right)\prod_{\mathfrak{P}\,|\,p}\left(1-\frac{\chi(N(\mathfrak{P}))\psi(\mathfrak{P})}{N(\mathfrak{P})^{s-k+2}}\right)^{-1}. \qquad (5.63)$$

Let us return now to the identity (5.59). On replacing the Dirichlet series on both sides of this identity by their product factorizations obtained above and dividing both sides by $\prod_{p\in\mathbb{P},\,p\nmid lq}(1-\chi(p^2)p^{-2(s-k+2)})$, we obtain to the relation (5.58). □

Let us reformulate the theorem in the terms of the multiplicativity of functions of integral argument. For an integer r, we say that a nonzero function

$$\varphi : \mathbb{N}_{\langle r\rangle} = \{m\in\mathbb{N} \,|\, \gcd(m,r)=1\} \mapsto \mathbb{C}$$

is $\langle r\rangle$*–multiplicative* if it satisfies the following two conditions:

(1) $\varphi(mm') = \varphi(m)\varphi(m')$ if m and m' are coprime (and prime to r);
(2) for each prime number p not dividing r, the formal power series $\sum_{v\geq 0}\varphi(p^v)v^v$ is formally equal to a rational fraction $N_p(v)D_p(v)^{-1}$, where $N_p(v)$ and $D_p(v)$ are polynomials with $N_p(0) = D_p(0) = 1$.

Alternatively, these conditions are equivalent to the condition that the following formal Euler factorization holds

$$\sum_{m\in\mathbb{N},\,\gcd(m,r)=1} \frac{\varphi(m)}{m^s} = \prod_{p\in\mathbb{P},\,p\nmid r} N_p(p^{-s})D_p(p^{-s})^{-1},$$

where $N_p(v)$ and $D_p(v)$ are polynomials with $N_p(0) = D_p(0) = 1$.

The function $m \mapsto \lambda(m)$, where $\lambda(m)$ are eigenvalues of a nonzero modular form $F \in \mathfrak{M}_k^2(q,\chi)$ satisfying (5.57), gives an example of $\langle q\rangle$-multiplicative functions. Other examples are given by the following corollary.

Corollary 5.29. *In the notation of Theorem 5.28, each of the functions $d \mapsto f(d\nu A)$, where A is a positive definite even primitive matrix of order two with discriminant $\delta = \mathbf{d}l^2$ and ν divides a power of lq, is a linear combination with constant coefficients of at most $h = h(\delta)$ $\langle lq\rangle$-multiplicative functions.*

Proof. It follows from Theorem 5.28 that each of the functions

$$d \mapsto \sum_{1\le i\le h} \psi(A_i) f(d\nu A_i)$$

is proportional to an $\langle lq\rangle$-multiplicative function. If, say, $A \sim A_j$, then

$$f(d\nu A) = f(d\nu A_j) = \frac{1}{h}\sum_{\psi} \overline{\psi}(A_j)\left(\sum_{1\le i\le h} \psi(A_i) f(d\nu A_i)\right),$$

by the orthogonality relations for the characters. □

Exercise 5.30. In the notation of Theorem 5.28 prove the following generalization of the relations (5.33):

$$\sum_{i=1}^{h} \psi(A_i) \sum_{\nu\in\mathbb{N},\,\nu\,|\,(lq)^\infty} f(\nu A_i) \prod_{p\in\mathbb{P},\,p\nmid lq} Q_{p,F}^{-1}(p^{-s})$$
$$= \prod_{\mathfrak{P}\nmid lq}\left(1 - \frac{\psi(\mathfrak{P})\chi(N(\mathfrak{P}))}{N(\mathfrak{P})^{s-k+2}}\right)^{-1} \sum_{i=1}^{h} \psi(A_i) \sum_{n=1}^{\infty} \frac{f(nA_i)}{d^s}.$$

Exercise 5.31. Prove that for each prime number p not dividing q, the polynomial (5.61) has a factorization of the form

$$Q_{p,F}(v) = (1-\alpha_0(p)v)(1-\alpha_0(p)\alpha_1(p)v)(1-\alpha_0(p)\alpha_2(p)v)(1-\alpha_0(p)\alpha_1(p)\alpha_2(p)v),$$

where $\alpha_0(p)^2\alpha_1(p)\alpha_2(p) = \chi(p^2)p^{2k-3}$.

[Hint: Use Proposition 3.55.]

5.3 Zeta Functions of Eigenforms for Genus 2

According to Theorem 4.7, each invariant subspace of cusp forms for the group $\Gamma_0^2(q)$ has a basis of eigenfunctions for all regular Hecke operators. The Zharkovskaya commutation relations (Theorem 4.19) allow one to prove that all regular Hecke operators can be simultaneously diagonalized on certain invariant spaces of modular forms consisting not only of cusp forms, but, for example, on the whole space $\mathfrak{M}_k^2$ of all modular forms of weight k for the full modular group Γ^2 (Theorem 4.22). This justifies a look at the eigenforms.

Regular Zeta Functions of Eigenforms. As in Section 5.2, we fix a nonzero modular form

$$F = F(Z) \in \mathfrak{M}_k^2(q,\chi) = \mathfrak{M}_k(\Gamma_o^2(q),\chi)$$

of positive integral weight k and character χ modulo $q \in \mathbb{N}$ for the group $\Gamma_0^2(q)$ and suppose that F is an eigenfunction for all Hecke operators $|T(m) = |_{k,\chi}T(m)$, where $T(m) \in \mathcal{L}_0(q) = \mathcal{L}_0^2(q)$ are the elements of the form (3.82), with the eigenvalues $\lambda(m)$,

$$F|T(m) = \lambda(m)F \qquad (\gcd(m,q) = 1).$$

It follows from Theorem 3.46 and Proposition 3.53 that one can write the formal identities

$$\begin{aligned}\sum_{m\in\mathbb{N},\,\gcd(m,q)=1} \frac{\lambda(m)}{m^s} &= \prod_{p\in\mathbb{P},\,p\nmid q}\sum_{\beta=0}^{\infty}\frac{\lambda(p^\beta)}{p^{\beta s}} \\ &= \prod_{p\in\mathbb{P},\,p\nmid q}\left(1-\frac{\chi(p^2)}{p^{2(s-k+2)}}\right)\prod_{p\in\mathbb{P},\,p\nmid q} Q_{p,F}^{-1}(p^{-s}), \qquad (5.64)\end{aligned}$$

where

$$\begin{aligned}Q_{p,F}(v) =& 1-\lambda(p)v + (\lambda(p)^2 - \lambda(p^2) - \chi(p^2)p^{2k-4})v^2 \\ &- \chi(p^2)p^{2k-3}\lambda(p)v^3 + \chi(p^4)p^{4k-6}v^4.\end{aligned}$$

We call the infinite product

$$\zeta_r(s,F) = \prod_{p\in\mathbb{P},\,p\nmid q} Q_{p,F}^{-1}(p^{-s}) \qquad (5.65)$$

the *regular zeta function of the eigenform* F; if $q=1$, we omit the adjective "regular" as well as the lower subscript r and get the *zeta function* $\zeta(s,F)$ *of* F.

Lemma 5.32. *The regular zeta function $\zeta_r(s,F)$ of a nonzero cusp form $F \in \mathfrak{N}_k^2(q,\chi)$ converges absolutely and uniformly in each right half-plane $\Re s > k+1+\varepsilon$ with $\varepsilon > 0$.*

Proof. Let $f(A)$ be a fixed nonzero Fourier coefficient of F. By relation (2.29) of Proposition 5.16 and estimate (1.46) of Proposition 1.25, for each m prime to q, we obtain the inequalities

$$|\lambda(m)| \leq \frac{c}{|f(A)|} \sum_{d,\, d_1,\, d_2 \in \mathbb{N},\, dd_1d_2=m} d_1^{k-2} d_2^{2k-3} \kappa(d_1) \det\left(\frac{d}{d_1d_2} A[\mathrm{diag}(1,\, d_1)]\right)^{k/2}$$
$$= \frac{c(\det A)^{k/2}}{|f(A)|} \sum_{d,\, d_1,\, d_2 \in \mathbb{N},\, dd_1d_2=m} d^k d_1^{k-2} d_2^{k-3} \kappa(d_1) < c' \tau_3(m) m^k,$$

where c is a constant, $\kappa(d_1) = \#(\Lambda_+ \backslash \Lambda_+ \mathrm{diag}(1, d_1)\Lambda_+) \leq d_1^2$ is the number of points on the projective line modulo d_1, and $\tau_3(m)$ is the number of factorizations of m into products of three positive integers. Hence using the elementary estimate $\tau_3(m) < c(\varepsilon')m^{\varepsilon'}$ with arbitrary $\varepsilon' > 0$, it follows that

$$|\lambda(m)| < c'' m^{k+\varepsilon'} \qquad (\gcd(m,\, q) = 1,\ \varepsilon' > 0), \tag{5.66}$$

where $c'' = c''(\varepsilon')$ is a constant depending only on ε'. It follows from the last estimate that the Dirichlet series on the left in (5.54) converges absolutely and uniformly on each of the indicated right half-planes. On the other hand, by (5.54), we can write the product (5.65) in the form

$$\zeta_r(s,\, F) = \left(\prod_{p\in\mathbb{P},\, p\nmid q} \left(1 - \frac{\chi(p^2)}{p^{2(s-k+2)}}\right)^{-1}\right) \left(\prod_{p\in\mathbb{P},\, p\nmid q} \sum_{\beta=0}^{\infty} \frac{\lambda(p^\beta)}{p^{\beta s}}\right)$$
$$= \left(\prod_{p\in\mathbb{P},\, p\nmid q} \sum_{\alpha=0}^{\infty} \frac{\chi(p^{2\alpha})}{p^{2\alpha(s-k+2)}}\right) \left(\prod_{p\in\mathbb{P},\, p\nmid q} \sum_{\beta=0}^{\infty} \frac{\lambda(p^\beta)}{p^{\beta s}}\right),$$

where absolute and uniform convergence of both products in the half-planes of the stated form follows from estimates (5.66) by a well-known classical test. For example,

$$\sum_{p,\, \beta\geq 1} \frac{|\lambda(p^\beta)|}{|p^{\beta s}|} < \sum_{m>1} \frac{|\lambda(m)|}{m^{\Re s}} < c'' \sum_{m>1} \frac{1}{m^{\Re s-k-\varepsilon'}} < c'' \sum_{m>1} \frac{1}{m^{1+\varepsilon-\varepsilon'}}$$

if $\Re s > k+1+\varepsilon$ with $\varepsilon > \varepsilon'$, and the last series with positive constant terms is convergent. □

It follows from the lemma that the regular zeta function $\zeta_r(s,\, F)$ of a cusp form $F \in \mathfrak{N}_k^2(q,\, \chi)$ is a holomorphic function of the complex variable s in each of the right half-planes $\Re s > k+1+\varepsilon$ with $\varepsilon > 0$. The question of analytic continuation of the zeta functions of cusp forms of level $q = 1$ will be considered in the next subsection.

Exercise 5.33. Prove that the regular zeta function of an arbitrary eigenform $F \in \mathfrak{M}_k^2(q,\chi)$ converges absolutely and uniformly in each right half-plane $\Re s > 2k+1+\varepsilon$ with $\varepsilon > 0$.

[Hint: Use the estimate (1.52).]

Exercise 5.34. Let $F \in \mathfrak{M}_k^2(q,\chi)$ be an eigenfunction for all regular Hecke operators. Prove that if $k \neq 2$ or the character χ is trivial, then the image $F|\Phi \in \mathfrak{M}_k^1(q,\chi)$ of F under the Siegel operator (1.56) is again an eigenfunction for all regular Hecke operators, and if $F|\Phi \neq 0$, the regular zeta functions of F and $F|\Phi$ are linked by the relation

$$\zeta_r(s,F) = \zeta_r(s,F|\Phi)\zeta_r(s-k+2,\chi;F|\Phi),$$

where

$$\zeta_r(s,\chi;F|\Phi) = \prod_{p\in\mathbb{P},\,p\nmid q} Q_{p,F|\Phi}^{-1}(\chi(p)p^{-s}).$$

Analytic Continuation and Functional Equation. Here we shall prove that under certain conditions, the zeta function of a cusp eigenform of level one has a meromorphic continuation over the whole s-plane and satisfies a functional equation with two gamma factors.

Theorem 5.35. *Let $F \in \mathfrak{N}_k^2$ be a cusp eigenform for all Hecke operators with Fourier coefficients $f(A)$. Suppose that $f\left(\left(\begin{smallmatrix}2&0\\0&2\end{smallmatrix}\right)\right) \neq 0$. Then the zeta function $\zeta(s,F)$ of F can be continued to the whole s-plane as a meromorphic function. More precisely, the function*

$$\Psi(s,F) = (2\pi)^{-2s}\Gamma(s)\Gamma(s-k+2)\zeta(s,F), \tag{5.67}$$

where $\Gamma(s)$ is the gamma function, has a continuation to the whole s-plane as a meromorphic function with at most two simple poles at the points $s = k-2$ and $s = k$ and satisfies the functional equation

$$\Psi(2k-2-s,F) = \Psi(s,F). \tag{5.68}$$

Proof. Let us use Theorem 5.28 in the case of the discriminant $\delta = \mathbf{d} = -4$ of the Gaussian number field $K = \mathbb{Q}(\sqrt{-1}) = \mathbb{Q}(\sqrt{-4})$ with class number $h = h(-4) = 1$, the representative $A_1 = 2\cdot 1_2$ of proper classes of discriminant -4, and the unit characters χ and ψ. Then the identity (5.60) takes the form

$$f(A_1)\zeta(s,F) = Z_{\mathcal{O}}(s-k+2)\sum_{d=1}^{\infty}\frac{f(dA_1)}{d^s}, \tag{5.69}$$

where

$$Z_{\mathcal{O}}(s) = \prod_{\mathfrak{P}}\left(1-\frac{1}{N(\mathfrak{P})^s}\right)^{-1} = \sum_{\mathfrak{A}}\frac{1}{N(\mathfrak{A})^s},$$

with $\mathfrak{P}$ and $\mathfrak{A}$ ranging over all prime ideals and nonzero integral ideals of the ring $\mathcal{O} = \mathbb{Q}[\sqrt{-1}]$ of the Gaussian integers, respectively, is the *Dedekind zeta function of the ring*. It follows from Lemma 5.32 and estimate (1.46) that both sides of (5.69) define functions holomorphic in the half-plane $\Re s > k+1$. Since $f(A_1) \neq 0$, in this half-plane we can write the functional identity

$$\begin{aligned}\Psi(s,F) &= (2\pi)^{-2s}\Gamma(s)\Gamma(s-k+2)\zeta(s,F)\\ &= f(A_1)^{-1}(2\pi)^{-2s}\Gamma(s)\Gamma(s-k+2)Z_{\mathcal{O}}(s-k+2)\sum_{d=1}^{\infty}\frac{f(dA_1)}{d^s}.\end{aligned}$$

According to Proposition 2.16, the function on the right has a meromorphic continuation to the whole s-plane with at most two simple poles at the points $s = k-2$ and $s = k$ and satisfies the functional equation (5.68). □

Exercise 5.36. Let $F \in \mathfrak{M}_k^2$ be an eigenfunction for all Hecke operators. Suppose that F is not a cusp form, i.e., $F' = f|\Phi \neq 0$. Prove that in this case, the function (5.67) is defined and holomorphic in the half-plane $\Re s > 2k+1$, has an analytic continuation to the whole s-plane as a holomorphic function if $F' \in \mathfrak{M}_k^1$ is a cusp form, and as a meromorphic function with four simple poles at the points $s = 0, k-2, k, 2k-2$, if F' is an Eisenstein series. In all cases it satisfies the functional equation (5.68).

Conclusion: Other Groups, Other Horizons

Although the theory of zeta functions of Siegel modular forms is far from being complete, and many fundamental questions (such as analogues of the Shimura–Taniyama conjecture for abelian varieties) have not even been touched, one must keep in mind that the symplectic group is but one example from a big family of arithmetically significant groups. The most natural situation, when it is possible to approach both analytic properties and factorization into an Euler product of arithmetic zeta functions, is the case in which zeta functions can be associated with representations of suitable arithmetic discrete subgroups of Lie groups on related function spaces of automorphic forms. The typical example of this kind is provided by "zeta functions of bilinear forms." Let, for example,

$$\mathbf{q} = \mathbf{q}(X,Y) = \sum_{i,j=1,\dots,m} q_{ij} x_i y_j = {}^t X Q Y \qquad \left(X = \begin{pmatrix} x_1 \\ \vdots \\ x_m \end{pmatrix}, Y = \begin{pmatrix} y_1 \\ \vdots \\ y_m \end{pmatrix} \right)$$

be a bilinear form of order m with integral nonsingular matrix $Q = (q_{ij})$. With the form $\mathbf{q}$ we associate the *automorph semigroup of* $\mathbf{q}$,

$$A(\mathbf{q}) = \left\{ D \in \mathbb{Z}_m^m \;\middle|\; \mathbf{q}(DX, DY) = \mu \mathbf{q}(X,Y) \quad \text{with } \mu = \mu(D) > 0 \right\},$$

and the *group of units of* $\mathbf{q}$,

$$E(\mathbf{q}) = \{ D \in A(\mathbf{q}) \mid \mu(D) = 1 \}.$$

The pair $(E(\mathbf{q}), A(\mathbf{q}))$ is quite often left-finite in the sense of Section 3.2, and we may define the Hecke–Shimura ring $\mathcal{D} = D(E(\mathbf{q}), A(\mathbf{q}))$ of the pair (over $\mathbb{Z}$), which is also called the *automorph class ring of* $\mathbf{q}$ and is denoted by $\mathcal{H}(\mathbf{q})$. Generally speaking, one has little to say about the ring $\mathcal{H}(\mathbf{q})$, and it should be replaced by the more complicated construction of a *matrix Hecke–Shimura ring*. Each bilinear form is the sum of uniquely defined *symmetric* and *skew-symmetric* forms, i.e., the forms $\mathbf{q}$ satisfying

$$\mathbf{q}(X,Y) = \mathbf{q}(Y,X) \quad \text{or} \quad \mathbf{q}(X,Y) = -\mathbf{q}(Y,X),$$

respectively, and corresponding groups and semigroups of automorphs are intersections of those for the components. In the case of either a pure symmetric or skew-symmetric form $\mathbf{q}$, each double coset of the semigroup $A(\mathbf{q})$ modulo the group $E(\mathbf{q})$ is a finite union of left cosets. In such a case, a theory can be developed that would resemble the Hecke–Shimura theory outlined above for the most interesting skew-symmetric case of the bilinear form with matrix $Q = J_n$ of order $2n$ defined in (0.1) (with the integral symplectic group $\Gamma^n = \mathrm{Sp}_n(\mathbb{Z})$ as the group of units and Siegel modular forms as representation spaces for symplectic Hecke–Shimura rings). In the symmetric case, when the form $\mathbf{q}$ is symmetric and the group of units $E(\mathbf{q})$ coincides with the proper integral orthogonal group of the quadratic form $\mathbf{q}(X,X)$, the theory of orthogonal matrix Hecke–Shimura rings is similar in many respects to the theory of symplectic rings, including the formal Euler factorization of generating Dirichlet series. In the orthogonal case, Hecke–Shimura rings operate on spaces of harmonic polynomials related to basic quadratic forms, and one can define zeta functions corresponding to these representations. It was revealed recently that in the case of positive definite quadratic forms in two and four variables, the relevant (orthogonal) zeta functions coincide with zeta functions corresponding to representations of symplectic Hecke–Shimura rings on spaces of theta series of genus 1 and 2 with harmonic coefficients, respectively. The coincidence is even more striking because Hecke–Shimura rings of quite different groups of units are involved: finite orthogonal groups and infinite symplectic groups. Other examples of relations between zeta functions of different arithmetic groups are provided by various lifting of automorphic forms and related zeta functions to similar groups of higher orders such as Saito–Kurokawa and Ikeda lifts, and various liftings of zeta functions of the general linear group to their symmetric degrees. There is no doubt that further progress in number theory will be closely connected with the investigation of relations among zeta functions of representations of Hecke–Shimura rings of various arithmetic discrete subgroups of Lie groups on automorphic functions.

Notes

Introduction

For the Riemann zeta function, see, e.g., [Ti86]. On zeta functions of algebraic varieties one can read [Shi71, Chapter 7]. For the Birch–Swinnerton-Dyer conjectures see [BSD63/65]. For modular forms in one variable and corresponding Dirichlet series see [Ogg69]. The Hecke theory of the Euler product factorization of Dirichlet series of modular forms is stated in [He37]. The original Atkin–Lehner theory was set forth in [AtL70]. For the history of the Shimura–Taniyama conjecture see [Shi89]. For the proof of Fermat's last theorem and related questions see [Wil95].

Chapter 1

The general references on this chapter, where one can find all omitted details and much more, are [An87, Chapters 1 and 2] and [AnZ90/95]. The later book also treats modular forms of half-integer weight. For a classical introduction to Siegel modular forms and Dirichlet series see [Kl90]. More details on the theory of Siegel modular forms and functions can be found in [Fr83], [Ma55], and [Si39]. For the treatment of automorphic forms on Lie groups see [SC58]. For modular forms in one variable see the books [Ogg69] and [Ma64]; for elementary presentations see [Ser70] and [Ga62]; relations to algebraic geometry are considered in [Fr83] and [Shi71]. Relations of modular forms to integral quadratic forms are considered in [Ki86] (numbers of representations of quadratic forms by quadratic forms) and in [An87] (multiplicative properties of the representations).

Section 1.2. For details on the reduction theory of positive definite quadratic forms see, e.g., [Ca78, Chapter 12]. For the construction of the fundamental domain of the modular group see also [Si73].

Section 1.3. The notation (1.29) is a variation of the notation introduced by H. Petersson. The Koecher effect was discovered in [Ko54/55]. For the rather complicated proof of the estimate (1.52) see, e.g., [An87, Theorem 2.3.4]. The scalar product of modular forms in one variable was introduced by H. Petersson in [Pe39/41]; the consideration of the general case is based on a similar idea and was initiated by H. Maass in [Ma51].

Chapter 2

Section 2.1. Radial Dirichlet series (2.10) are not the only kind of Dirichlet series constructed with the help of Fourier coefficients of common eigenfunctions for regular Hecke operators that have an Euler product factorization and good analytic properties. Another kind of such Dirichlet series is given by series of the form

$$\sum_{M \in \mathrm{SL}_n(\mathbb{Z})\backslash\{M \in \mathbb{Z}_n^n \mid \det M > 0\}} \frac{\psi(\det M) f(MA{}^tM)}{(\det M)^s},$$

where $n \geq 1$, f are Fourier coefficients of a common eigenfunction for regular Hecke operators on the space $\mathfrak{M}_k(\Gamma_0^n(q), \chi)$ with integral or half-integral k, ψ is a Dirichlet character, and A is a fixed even positive definite matrix of order n. For $n = 1$ these series were considered in [Ra39], [Se40], and [Shi75]; Euler product factorizations for an arbitrary n and integral k were discussed in [An87, §4.3.3], whereas both integral and half-integral k were considered in [AnZ90/95, §3.3]; analytic properties of corresponding Euler products (the *standard zeta functions of eigenforms F*) of level $q = 1$ with $\psi = 1$ were considered in [AnK78] with certain restrictions, and in [Bo85] without restrictions.

Section 2.3. Our exposition of the theory of radial Dirichlet series corresponding to the ray $mA_1 = 2m \cdot 1_2 (m = 1, 2, \ldots)$ of the matrix of a quadratic form $x_1^2 + x_2^2$ was outlined first in [An71, §3]; the general case of radial series corresponding to rays of matrices of arbitrary positive definite integral primitive binary quadratic forms was considered in detail in [An74, §§3.3–3.7].

On Chapter 3

Section 3.1. The multiplicative properties of the Fourier coefficients of the modular form $\Delta'(z) \in \mathfrak{N}_{12}(\Gamma)$ cited in Exercise 3.1(3) were observed (!) and conjectured by Ramanujan and a little later proved by Mordell in [Mo17]. Mordell's proof actually contains the idea of Hecke operators for the particular case of the space $\mathfrak{N}_{12}(\Gamma)$, an idea that had to wait twenty more years to be reborn in the case of more general spaces of modular forms by Hecke in [He37].

Section 3.2. Our definition of an abstract ring of double cosets by means of multiplication of right-invariant linear combinations of left cosets is equivalent to the definition given by Shimura in [Shi63] with the help of the direct multiplication of double cosets, but in some respects it turns out to be more convenient.

Section 3.3. The rings of double cosets of the general linear group are discussed here and in [An87, §3.2] in the spirit of the fundamental paper [Ta63] of Tamagawa. For the explicit formulas for spherical functions on GL_n see [An70].

Section 3.4. The symplectic case is described as in [An87, §3.3]. The structure of Hecke–Shimura rings for Γ^n was indicated first in [Sa63] and [Shi63]. I cannot judge who was the first. The passage to congruence subgroups goes back to Hecke's ideas in [He37]. Our presentation uses common sense and an analogy with the case of the general linear group rather than any historical reminiscences. For the theory of singular Hecke–Shimura rings see [An99]. The theory of spherical mappings in the form of a general theory of zonal spherical functions on reductive algebraic groups over $\mathfrak{p}$-adic fields is due to Satake; see [Sa63]. We use an elementary approach based on explicit formulas. The series (3.107) and (3.109) were summed up in [Shi63]; similar series for arbitrary n were computed in [An69] and [An70].

Section 3.5. The embedding of symplectic Hecke–Shimura rings into rings of triangular-symplectic double cosets allows one to split their elements into elementary components naturally related to the general linear group. This gives impetus to the theory of factorization of standard symplectic polynomials in triangular extensions, presented in detail in [An87, §§3.4–3.5]. Here we use only the elementary aspects of the theory.

Chapter 4

General references are [An87, §§4.1–4.2] or [AnZ90/95].

Section 4.1. Hecke operators for the group $\Gamma^1 = SL_2(\mathbb{Z})$ and some congruence subgroups were introduced in [He37]. Hecke operators on Siegel modular forms were first introduced by Sugawara in [Su37] and [Su38] and after the war were studied by Maass in [Ma51]. The existence of a basis of common eigenfunctions for Hecke operators acting on cusp forms in one variable was proved by Petersson in [Pe39/41]. For Siegel modular forms, Petersson's ideas were developed by Maass in [Ma51]. For diagonalization of singular Hecke operators see [AtL70] in the case of genus $n = 1$, and [An99], [An03] in the general case.

Section 4.3. Zharkovskaya commutation relations for the group Γ^n and the unit character are due to Natasha Zharkovskaya [Zha74], who generalized the Maass relation for Γ^2 obtained in [Ma51]. Our consideration follows the same idea.

Chapter 5

Section 5.1. Relations of Fourier coefficients of eigenforms with eigenvalues can be found in [He37].

Section 5.2. For more details on the relations of binary quadratic forms and modules in quadratic number fields see, e.g., [An87, Appendix 3], or the excellent book on algebraic numbers [BSh63, Chapter 2].

Section 5.3. It is proved in [An74, Theorem 3.1.1] without any restrictions that the zeta function $\zeta(s, F)$ of an arbitrary eigenfunction for all Hecke operators $F \in \mathfrak{M}_k^2$ of integral weight $k \geq 0$ can be continued to the whole s-plane as a meromorphic function; the function

$$\Psi(s, F) = (2\pi)^{-2s}\Gamma(s)\Gamma(s-k+2)\zeta(s, F),$$

where Γ is the gamma function, is meromorphic on the s-plane with the only possible poles at $s = 0, k-2, k, 2k-2$ and satisfies the functional equation

$$\Psi(2k-2-s, F) = (-1)^k\Psi(s, F).$$

This does not contradict the functional equation (5.67), because the assumptions of Theorem 5.35 imply that k there must be even.

The relation between the eigenvalues of Hecke operators and Fourier coefficients of eigenfunctions for Siegel modular forms of genus n discovered by Zharkovskaya in [Zha75] can possibly provide the first step to approach the still open problem of analytic properties of similar zeta functions of cusp form for genera $n > 2$.

Conclusion

On relations of zeta functions of orthogonal and symplectic groups see [An06]. For a general outlook on problems of Euler products attached to automorphic forms see Langlands' lectures [La67] and [La69].

References

[An69] A.N. Andrianov, *Rationality theorems for Hecke series and zeta-functions of the groups* GL_n *and* Sp_n *over local fields*, Izv. Akad. Nauk Ser. Mat. **33** (1969), 466–505 (Russian); English transl., Math. USSR Izv. **3** (1969), 439–476.

[An70] A.N. Andrianov, *Spherical functions on* GL_n *over local fields and summation of Hecke series*, Mat. Sbornik Nov. Ser. **83** (1970), no. 3, 429–451 (Russian); English transl., Math. USSR Sb. **12** (197), no. 3, 429–452.

[An71] A.N. Andrianov, *Dirichlet series with Euler product in the theory of Siegel modular forms of genus* 2, Trudy Mat. Inst. Steklov **112** (1971), 73–94 (Russian); English transl., Proc. Steklov Inst. Math. **112** (1971), 70–93.

[An74] A.N. Andrianov, *Euler products corresponding to Siegel modular forms of genus* 2, Uspekhi Mat. Nauk **29** (1974), no. 3, 43–110 (Russian); English transl., Russian Math. Surveys **29** (1974), no. 3, 45–116.

[An87] A.N. Andrianov, *Quadratic forms and Hecke operators*, Grundlehren Math. Wiss. 286, Springer-Verlag, Berlin Heidelberg New York London Paris Tokyo, 1987.

[An99] A.N. Andrianov, *Singular Hecke–Shimura rings and Hecke operators on Siegel modular forms*, Algebra i Analis **11** (1999), no. 6, 1–68; English transl., St. Petersburg Math. J. **11** (2000), no. 6, 931–987.

[An03] A. Andrianov, *On diagonalization of singular Frobenius operators on Siegel modular forms*, Amer. J. Math. **125** (2003), 139–165.

[An06] A.N. Andrianov, *Zeta functions of orthogonal groups of integral positive definite quadratic forms*, Uspekhi Mat. Nauk **61** (2006), no. 6, 3–44 (Russian); English transl., Russian Math. Surveys **61** (2006), no. 6.

[AnK78] A.N. Andrianov and V.L. Kalinin, *On the analytical properties of standard zeta functions of Siegel modular forms*, Mat. Sbornik **106(148)** (1978), no. 3, 323–339 (Russian); English transl., Math. USSR Sbornik **35**, no. 1, 1–17.

[AnZ90/95] A.N. Andrianov and V.G. Zhuravlev, *Modular forms and Hecke operators*, "Nauka", Moscow, 1990 (Russian); English transl., Transl. of Math. Monographs, vol. **145**, AMS, Providence Rhode Island, 1995.

[AtL70] A.O.L. Atkin and J. Lehner, *Hecke operators on* $\Gamma_0(m)$ Math. Ann. **185** (1970), 134–160.

[BSD63/65] B.J. Birch and H.P.F. Swinnerton-Dyer, *Notes on elliptic curves*. I,II, J. Reine Angew. Math. **212** (1963), 7–25; **218** (1965), 79–108.

[Bo85] S. Böcherer, *Über die Funktionalgleichung automorpher L-Funktionen zur Siegelschen Modulgruppe*, J. Reine Angew. Math. **362** (1985), 146–168.

[BSh63] Z.I. Borevich and I.R. Shafarevich, *Number Theory*, "Nauka", Moscow, 1963, 1971, (1985) (Russian); English transl., Academic Press, New York, 1966.

[Fr83] E. Freitag, *Siegelsche Modulfunktionen*, Grundlehren Math. Wiss. 254, Springer-Verlag, Berlin Heidelberg New York London Paris Tokyo, 1983.

[Ga62] R.C. Gunning, *Lectures on modular forms*, Annals of Math. Studies 48, Princeton University Press, Princeton, New Jersey, 1962.

[He37] E. Hecke, *Über Modulfunktionen und die Dirichletschen Reihen mit Eulerscher Produktentwicklung*. I, II, Math. Ann. **114** (1937), 1–28; 316–351; Math. Werke, Vandenhoeck & Ruprecht, Göttingen, 1959, 1970, pp. 644–707.

[Ki86] Y. Kitaoka, *Lectures on Siegel modular forms and representations by quadratic forms*, Published for the Tata Institute of Fundamental Research, Springer-Verlag, Berlin Heidelberg New York Tokyo, 1986.

[Kl90] H. Klingen, *Introductory lectures on Siegel modular forms*, Series: Cambridge Studies in Advanced Mathematics (No 20), Cambridge University Press, 1990.

[Ko54/55] M. Koecher, *Zur Theorie der Modulformen n-ten Grades*. I, II, Math. Z. **59** (1954), no. 4, pp. 399–416; **61** (1955), no. 4, 455–466.

[La67] R.P. Langlands, *Euler products*, Lecture Notes, Yale University, New Haven, Connecticut, 1967.

[La69] R.P. Langlands, *Problems in the theory of automorphic forms*, Lecture Notes, Yale University, New Haven, Connecticut, 1969.

[Ma51] H. Maass, *Die Primzahlen in der Theorie der Siegelschen Modulfunktionen*, Math. Ann. **124** (1951), 87–122.

[Ma55] H. Maass, *Lectures on Siegel's modular functions*, Tata Inst. Fund. Research, Bombay, 1955.

[Ma64] H. Maass, *Lectures on modular functions of one complex variable*, Revised 1983, Tata Inst. Fund. Research, Bombay, 1964.

[Mo17] L.J. Mordell, *On Mr Ramanujan's empirical expansions of modular functions*, Proc. Cambridge Phil. Soc. **19** (1917), 117–124.

[Ogg69] A.P. Ogg, *Modular forms and Dirichlet series*, W.A. Benjamin, Inc., New York, Amsterdam, 1969.

[Pe39/41] H. Petersson, *Konstruktion der sämtlichen Lösungen einer Riemannschen Funktionalgleichung durch Dirichletreihen mit Eulerscher Produktentwicklung*. I, II, III, Math. Ann. **116** (1939), 401–412; **117** (1939), 39–64; **117** (1940/41), 277–300.

[Ra39] R.A. Rankin, *Contributions to the theory of Ramanujan's function $\tau(n)$ and similar arithmetical functions*. II, The order of the Fourier coefficients of the integral modular forms, Proc. Cambridge Phil. Soc. **35** (1939), no. 3, 357–372.

[Se40] A. Selberg, *Bemerkungen über eine Dirichletsche Reihe, die mit der Theorie der Modulformen nahe verbunden ist*, Arch. Math. Naturvid. **43** (1940), 47–50.

[SC58] Sèminaire H. Cartan, *Fonctions automorphes*, 10e année, vol. 1, 2, Secrètariat Mathématique, Paris, 1958.

[Ser70] J.-P. Serre, *Cours d'arithmétique*, Presses Universitaires de France, Paris, 1970.

[Shi63] G. Shimura, *On modular correspondences for* $\mathrm{Sp}(n,\mathbb{Z})$ *and their congruence relations*, Proc. Nat. Acad. Sci. USA **49** (1963), no. 6, 824–828.

[Shi75] G. Shimura, *On the holomorphy of certain Dirichlet series*, Proc. London Math. Soc. **31** (1975), no. 1, 79–98.

[Shi71] G. Shimura, *Introduction to the arithmetic theory of automorphic functions*, Publ. Math. Soc. Japon, vol. 11, Iwanami Shoten, Publishers and Princeton University Press, Princeton, 1971.

[Shi89] G. Shimura, *Yutaka Taniyama and his time*, Bull. London Math. Soc. **21** (1989), 186–196.

[Si39] C.L. Siegel, *Einführung in die Theorie der Modulfunktionen n-ten Grades*, Math. Ann. **116** (1939), 617-657; Gesam. Abhandlungen vol. II, pp. 97–137.

[Si73] C.L. Siegel, *Topics in complex function theory*, Interscience Tracts in Pure and Applied Math. 25 vol. III, Wiley-Interscience, New York London Sydney Toronto, 1973.

[Ta63] T. Tamagawa, *On the ζ-functions of a division algebra* Ann. Math. II Ser. **77** (1963), no. 2, 387–405.

[Tit86] E. C. Titchmarsh, *The theory of the Riemann zeta-function*, 2nd ed., Clarendon press, Oxford, 1986.

[Wil95] A. Wiles, *Modular elliptic curves and Fermat's Last Theorem*, Ann. Math. II Ser. **141** (1995), no. 3, 443–551.

[Zha74] N.A. Zharkovskaya, *The Siegel operator and Hecke operators*, Funkts. Anal. Prilozh. **8** (1974), no. 12, 30–38; English transl., Funct. Anal. Appl. **8** (74), 113–120.

[Zha75] N.A. Zharkovskaya, *On the connection between the eigenvalues of Hecke operators and Fourier coefficients of eigenfunctions for Siegel's modular forms of genus n*, Mat. Sb. Nov. Ser. vol 96(138) (1975) 584–593; English transl., Mat. USSR Sb. **25** (1975).

Index